MW00379001

INTO THE
UNKNOWN

INTO THE UNKNOWN

THE QUEST TO UNDERSTAND THE MYSTERIES OF THE COSMOS

KELSEY JOHNSON

BASIC BOOKS
New York

Basic Books
Hachette Book Group
1290 Avenue of the Americas, New York, NY 10104
www.basicbooks.com

Printed in the United States of America

First Edition: October 2024

Published by Basic Books, an imprint of Hachette Book Group, Inc. The Basic Books name and logo is a registered trademark of the Hachette Book Group.

The Hachette Speakers Bureau provides a wide range of authors for speaking events. To find out more, go to hachettespeakersbureau.com or email HachetteSpeakers@hbgusa.com.

Basic books may be purchased in bulk for business, educational, or promotional use. For more information, please contact your local bookseller or the Hachette Book Group Special Markets Department at special.markets@hbgusa.com.

The publisher is not responsible for websites (or their content) that are not owned by the publisher.

Print book interior design by Bart Dawson.

Library of Congress Cataloging-in-Publication Data

Names: Johnson, Kelsey, author.
Title: Into the unknown : the quest to understand the mysteries of the
    cosmos / Kelsey Johnson.
Description: First edition. | New York : Basic Books, 2024. | Includes
    bibliographical references and index.
Identifiers: LCCN 2024002952 | ISBN 9781541604360 (hardcover) |
    ISBN 9781541604384 (ebook)
Subjects: LCSH: Cosmology—Popular works.
Classification: LCC QB982 .J64 2024 | DDC 523.1—dc23/eng/20240509
LC record available at https://lccn.loc.gov/2024002952

ISBNs: 9781541604360 (hardcover), 9781541604384 (ebook)

LSC-C

Printing 1, 2024

To all who gaze at the night sky and wonder

CONTENTS

Contents

Prologue

ONE DAY WHILE WALKING HOME FROM ELEMENTARY SCHOOL, two apparently contradictory facts came together in my head at the same time, and I was both confused and mad.

"So, if fire needs air to burn, and there is no air in space, how do stars burn?"

Either I was too stupid to understand how these "facts" could make sense together, or one of these things had to be wrong, which meant the universe didn't make sense. I asked my mom this question later that night, and she responded, "I don't know, I'm not smart enough. I wish your dad were around, I bet he could answer it." (My mom *was* smart, but she didn't have an education past high school.) That night I knew there were things about the universe that were a mystery to me. I also knew that I had to try to understand them. At the time, I also believed that everything must ultimately be understandable, even if I didn't understand it.

In middle school my science teacher (thanks, Mr. Sackett!) took our class to visit the local state university. We sat in on several classes, which included an astronomy lecture on asteroids causing the extinction of dinosaurs. I'd never really thought about why the dinosaurs went extinct. I just took it at face value as "a thing that had happened that maybe someone understood." I was shocked that no one had ever told me about asteroids killing dinosaurs. We also sat in on a physics class that day, during which the professor did amazing demonstrations of angular momentum and magnetism,

1

and I realized there were real things in the physical world that I could not see that were extremely powerful and tantalizingly mysterious to me. I didn't fully know what it meant, but I came home from school that day and declared to my mom that I was going to be an astrophysicist.

In high school, one of my teachers took us to the library at the nearby state university, which made my local library seem quaint and provincial by comparison. I literally had no idea that so much was known about so many things. Maybe here I could find answers to the mysteries of the universe. I checked out books from the university library on black holes, time travel, and other dimensions. I couldn't understand them and the strange symbols and math they used, but that just made the mysteries seem even more enticing.

When I got to college, I was primed to learn everything I could possibly learn about the universe. Sheer momentum kept me on my quest for understanding for a long time, but after several semesters of learning the detailed math and physics necessary to understand the universe, it became clear to me that the smallest details took center stage. Shockingly few professors engaged with the deep mysteries of the universe that I thought the course material was supposed to be preparing me for. While it may be hard to "see the forest for the trees," it is at least equally hard to see the cosmos for the stars.

At the time, I assumed that all my professors knew the answers to the big questions I had, but I just didn't have enough background knowledge yet to understand if they tried to share the answers with me. In hindsight, I think they avoided talking about the big questions because they didn't know the answers either, instead keeping their focus on the details and challenges of their particular studies: How massive is that black hole? Will those galaxies collide? When will the next supernova explode? These are all good and valuable questions in our quest to understand the universe, but they are a step removed from the beautiful existential questions that impel so many of us to study astrophysics to begin with.

The more I learned, the more I realized just how limited our understanding of the cosmos is. So many fundamental questions remained: What caused the Big Bang? What is the nature of time? What is dark matter? And not least of all, what is knowledge? I packed up my quest for real understanding and tucked it away for later—perhaps after I got my PhD, I could unpack it again with some hope of making progress.

After ten years of college and graduate school, and now many years of teaching at one of the top public universities in the world, I have honed my understanding of what is known and what may be unknowable. Through transdisciplinary work, particularly with colleagues in religious studies, I have also gained valuable experience and insight into diverse epistemological perspectives.

Over these years, I've developed a passion for trying to help people understand the foundations of the great mysteries of the cosmos—without needing years of math and science. Yes, of course, knowing advanced math and science will help your intuition, but I believe there is a lot you can understand without a PhD in astrophysics. I just have this crazy belief that it would do humanity good if we all had a better understanding of the universe we are part of.

If you are hoping for a good read on "solved mysteries," you may wish to look elsewhere. There is no shortage of mysteries in the universe, which is great job security for scientists like me, but presents a bit of a challenge when deciding what makes the cut for a book. I did what any normal human might do and picked my absolute favorites to think about; these are the topics that keep me ruminating well past normal sleeping hours.

The quandaries we will encounter in this book reside at the boundary of human knowledge and understanding. The unknown—and potentially unknowable—nature of each topic sits at the convergence of science, philosophy, and theology. We will encounter question after question to which the answer could be "God." To be clear at the outset, I don't dismiss the possibility of a higher power of some sort. In fact, my exposure to astrophysics and the unsolved mysteries of the universe has led me to conclude through purely logical means that there must be things about the universe and its origin that we don't understand, and possibly never will. Is there room for a higher power? Sure. However, I am also highly sensitized to the brainwashing that takes place in our society, and the extent to which people are taught not to think for themselves. Any belief system too fragile to permit people to reach their own conclusions is inherently flawed and unstable. When we hit the limits of our understanding in each chapter of this book, I will do my level best to offer a buffet of possible solutions, many of which are not mutually exclusive. A higher power will often be among these solutions, as will "something we haven't thought of yet." But I'm not going to tell you the "right" answers, because we humans do not yet have these at our disposal.

The topics of this book are entwined in complex ways—it turns out the universe didn't evolve to have nice, tidy, and separate categories for humans to write nice, tidy books with clear-cut chapters. As you go on this journey with me through these baffling topics, the deeper we get into the landscape of the mysteries, the more actively connections between topics will multiply. If you are so inclined, you might even consider reading this book *twice*—building your knowledge of the landscape on the first pass and making deeper connections between chapters on the next go round. Then again, finding time to read at all can be a luxury, and I'm grateful for your curiosity about the universe, whether you just read a single chapter, or the whole book. Twice.

I hope your journey through this book will inspire you to think—for yourself—about the great mysteries of the universe, and by the last page you are even more perplexed and curious about the cosmos we are lucky enough to find ourselves in.

1

A Little Perspective

OFTEN WHEN WE ASTRONOMERS MENTION THAT WE ARE astronomers (or "astrophysicists" if we are feeling antisocial), we get a response along the lines of,

"Cool! I loved learning constellations. But don't we already pretty much know everything?"

The answer is absolutely, conclusively, no.

We've barely scratched the tip of a single ice crystal on the tip of the iceberg. In fact, the iceberg itself may even be an illusion. One of the (many) reasons I have trouble at parties is that small talk and the actual nature of reality don't play well together.

You may go about your normal life and feel things just *work*. The microwave heats your frozen meal (if not uniformly). Cars drive (and are even starting to drive themselves). Computers calculate all kinds of things—some of which are even useful. We've sent spacecraft all over the solar system, and we're on the verge of sending actual living people on similar escapades to places like Mars. To be sure, we humans have learned a lot, and we seem to have more-or-less understood the world of our everyday lives. So, you can be forgiven if you think we already know pretty much everything. But I'm writing this book with the hope of dislodging that comfy notion from your brain. Considering how limited and provincial our experiences are, I find it astounding that we've even figured out as much as we have about the universe.

One of the reasons our world seems to make sense is simply that we are used to it. It is easy to go about normal daily life and take basic things for granted. *They just are.* For example, right now you may be sitting on a chair, or standing on the floor (or sitting on the floor—but hopefully not standing on a chair). That's just about the most everyday kind of experience you might have. Take a generous moment and think about gravity holding you down, literally pulling your mass toward the mass of the Earth. *Why* does gravity do that? "Just because it does" is not a very satisfying answer, and if you've had the pleasure of interacting with kids, they might let you know just how unsatisfying "Just because" is, as far as answers go. While we're on the topic, what even *is* mass? To be fair, we do know a bit about what is under the hood as far as mass and gravity go, but just because I know there is an engine under the hood of a car, that doesn't mean I know how it works (which is factually true in my case).

Coming to terms with our unfathomable insignificance in the cosmos may be impossible—the scale of the universe in space and time extends so far beyond our ability to comprehend that sometimes I feel the best we can do is try to comprehend that we don't comprehend. Even for astronomers who think about these concepts daily, the full scales of time and space are largely abstractions many of us have become desensitized to—perhaps out of necessity to keep our sanity. Staring into the abyss every single day can take a toll if one doesn't get a touch habituated to it.

Yet here we are on this tiny little speck of a planet trying to understand not just *this* tiny speck we call home, but the *whole universe.* It is a monumental task to try to understand the cosmos, and there is no doubt that we have barely scratched the surface. The fact that we have sent relatively infinitesimally tiny little tin cans of space-crafts successfully to all the other specks we call planets in the solar system and gotten data back from them is an astounding testament to human ingenuity. So, despite our mind-blowing insignificance in the cosmos, you can take a beat and be a little proud of humanity (but not too proud—we've screwed up a lot of stuff, too).

To be clear, we only have a limited set of tools at our disposal. Astronomers are effectively scavengers: aside from a few crumbs of information we gather from within the solar system, pretty much the only information we receive from the universe comes from light. So, we've gotten *really* good at scavenging light and squeezing out every last bit of information it can give us. If you were to decide to become an astronomer, you would spend a lot of time learning how to analyze light. Other than light,[1] we are largely limited to what we can do with our own brains and logic.

Still, I think it is essential to try to convey at least an impression of the universe we live in. The catch is that this is hard to do on the scale of a book—presumably the book you are reading is smallish on the scale of the universe. Just by seeing things in this smallish book, your brain does a neat trick when you see, for example, a picture of a galaxy. Yup, the galaxy fits on the page, which is totally in the realm of "normal." This is a very different experience than, say, standing (or more likely floating) on the edge of a galaxy and trying to compre-hend its vastness extending in every direction as far as you can see.

Humanity has now taken some astounding images of the uni-verse. We even have images recording light that left its source almost 13 billion years ago and traveled across the entire visible universe before hitting one of our telescopes.[2] In other words, we can look back in time almost 13 billion years. How does that make you feel? Like you're standing on the edge of a dark and unfathomably deep abyss? Good. Because you are.

To provide even an inkling of a sense of scale, let's start with something closer to home. If you could take a road trip to the Moon on an imaginary superhighway that still had normal Earthly speed limits, how long do you think it would take? Answer: about half a year (without stopping, so maybe stock up on diapers). But you can at least fathom half of a year—that is a timescale you have experienced many times unless you are an exceptionally young and precocious reader. As a reality check, the Moon is the farthest we have ever sent people. To be clear, this is not necessarily a bad thing. I am not a fan of unfettered human colonization. Our ethical understanding of the implications is alarmingly far behind current corporate aspirations. Still, to date we have only managed to get to the moon a couple times in the middle of last century. That's it.

The farthest that we routinely have people living today is the International Space Station, which might understandably give you the idea that we humans are actually going to space. If you could drive to the space station when it is overhead it would take something like a mere four hours. Leave after breakfast and get there for lunch (hypothetically).

We are not going to zoom out to the entire scale of the universe, but I want to take one tiny step out to the scale of the solar system to make a point. If the Sun were the size of a grapefruit, the Earth would be roughly the size of one of those little round sprinkles you might put on cupcakes. Imagine holding them up next to each other for comparison (or if you have grapefruit and sprinkles on hand, you could actually hold them up for comparison and invite inquisitive looks from your family, roommates, or cat). As a fun fact, you could fit about a million Earths inside the Sun (if you were so inclined, had superpowers, and didn't care about getting toasted). My experience is that the true relative scale of the Earth and the Sun is radically underappreciated, and for understandable reasons; to illustrate the solar system, many books don't show things to scale—in part because they *can't* show things to scale if you want to see anything. The standard presentation of not-to-scale images can leave an unwary reader with the misimpression that the Earth is a heck of a lot bigger and closer to the Sun than it is. If this is the visualization we grow up with engrained in our minds as reality, our opinion of ourselves and our place in the universe is surely distorted.

On this grapefruit scale, we can now think about the true distances to things like other stars. If the Sun-grapefruit were in Washington, DC, the nearest star-grapefruit would be roughly somewhere on the West Coast of the US, say Seattle or Los Angeles. That is a lot of space between grapefruits. The metapoint is, of course, the universe is really big. This is an obvious statement that I think has lost its bite, at least in part because it is virtually impossible for us mere humans to comprehend the scales involved. As a result, instead of confronting the magnitude of these scales (and our apparent relative insignificance), we are inclined to put this idea in a box and tuck it under the bed where we don't need to think about it. This may be a well-honed defense mechanism to avoid a state of continual existential crisis, and I get that. But I am of the opinion that at least a little existential crisis is good for the soul.

Existential experiences that induce us to reflect on the nature of being and the limits of our knowledge are rich sources of awe. A body of research from recent decades has shown us the importance of experiencing awe in human flourishing. Among many other benefits, experiencing awe enhances creativity, curiosity, and critical thinking.[3] Psychological work has distilled the experience of awe as a confluence of "vastness and accommodation"; "vastness" refers to anything that is "experienced as being much larger than the self, or the self's ordinary level of experience or frame of reference."[4] "Accommodation" refers to the need to modify existing mental structures to assimilate this new information or experience.[5]

Surely, it is hard to rival exposure to the universe in either category of "vastness" or "need for accommodation." The sheer fact of the existence of the cosmos and that we have "something" instead of "nothing" may be a reality and a concept beyond human comprehension. When we add in the existence of things like an invisible force so powerful that it is causing the universe to accelerate, the fact that time is malleable, that the dimensions we perceive may only be a shadow of reality, or even the humble truth of Euler's equation, you may be able to almost feel your brain stretching to accommodate these concepts that extend so far beyond human experience.

Don't worry, we're not going to immediately just pull out the box of existential issues and rip the lid off—we have a little preparation work to do first. Before we can meaningfully talk about what we *don't* know, we need to spend some quality time thinking about how (or if) we know things at all, and what the role of science, theology, and philosophy are in this endeavor.

Forget What You Think You Know About Science

For many people, their exposure to science is limited to classes in high school and maybe a couple huge introductory science courses if they went to college. Students often leave these experiences with the impression that science is all about memorizing facts, knowing how to solve equations, and doing "experiments" that literally *millions of other people* have done before and to which there is a "right" answer. The essential nature of science is completely lost in experiences like this, which—to be clear—have virtually nothing in common with doing actual science. At its core, science is about playing with stuff to uncover new things about the universe (which, by the way, includes our planet and everything on it) that are brand-new to you—and maybe brand-new to anyone.

By analogy—think about spelling and grammar. There are people who enjoy learning how to spell and use correct grammar. I hypothesize that people who love learning how to spell are somewhat rare, and most people just put up with learning these things because they are told they have to. Still, many students learn the fundamentals of spelling and grammar before high school and then go on to do *actually interesting* things (from my perspective), like reading great novels or writing poetry (and I think tend to forget how annoying it was to learn the building blocks to begin with). But you would be hard-pressed to read a great novel or write a poem without the fundamental skills of spelling and grammar in your tool kit.

When it comes to science, memorizing facts, knowing how to solve equations, and doing so-called "experiments" that are in no danger of uncovering anything new are the equivalent of learning spelling and grammar. When it comes to science, many people (through no fault of their own) never get to the point where they can do the science equivalent of reading a great novel or writing a poem. An enormous challenge in the education system is to find ways to expose students to the joy (yes, actual bona fide joy) that can accompany doing real science—by which I mean an experiment you have designed to answer a question you actually want to know the answer to, and to which no one might actually know the answer. Sometimes scientists just do this for fun.

Many people also have sentiments about science and scientists that lean deeply into stereotypes—after all, how many professional scientists does a typical person know and interact with regularly? For example, there is an irony in people thinking so uncreatively about creativity to think that science doesn't require it. If you want to solve big outstanding mysteries (or even small ones), you must be able to come up with new ideas, some of which will seem crazy. Following the rules in science is basically the equivalent of learning the rules for spelling and grammar. Do you need to follow rules when writing a novel? Well, sort of. There are elements of writing that are best practice, and if you don't use them, your work may be unintelligible to the outside world (thank goodness I have an editor). But if all you did was follow formulaic rules, your novel would probably not be a big hit. The same is true for science; there are elements of doing science that are important to making sure your results are robust and usable to the broader community. But if you want to make a breakthrough, creativity is essential.

There is also a pervasive opinion that there is little room for interpretation in science. Friedrich Nietzsche is often quoted as saying, "All things are subject to interpretation. Whichever interpretation prevails at a given time is a function of power and not truth" (although I note with some irony that this is not actually what he said).[6] In science, we collect data, and then try to assess what theory best fits those data. That sounds straightforward enough, but deciding which theory fits best depends on what elements you think are most important to fit, whether you agree with assumptions that have been made, and whether you believe the data are sufficiently representative, and so on. Then the question you must ask (as a scientist) is "How can I test whether my interpretation is correct?" If you can't test it, you may want to dial back your confidence that you're right.

That is, we also need to be skeptical. The word "skeptical" has been hijacked by colloquial language. For scientists, "skeptical" does *not* mean that you don't believe anything. "Skeptical" *does* mean that you look at the strengths and weaknesses of any given claim and prefer to have evidence. In the latter sense of the word "skeptical," scientists are (or should be) skeptical. In my opinion, so should everyone else. If you believe everything anyone tells you with no need for evidence, then I'd like to sell you some stars in the night sky (which, just FYI, is a total scam—sorry). Do you believe every news source you read without question? How about politicians? Also, to be clear, it isn't actually possible to believe everything you are told because inevitably you will be told things that conflict. How do you know which to believe? Skepticism to the rescue.

Finally, I am all about using science as a tool to help us understand the universe, but if you want to use a tool most effectively, it is helpful to also understand its limitations. To be sure, there are occasions in which I have used the handle of a screwdriver as a hammer, or a table knife as a screwdriver, but the result would have been better if I'd had a more suitable tool at hand. As far as science goes, it is really good at testing things that are testable, but outside of the realm of the testable, science has no purchase. And this is the very realm where many of the most profound questions about the cosmos dwell.

The Boundaries of Science

We can do, and have done, an impressive amount with our brains and logic. But there are limits. Sometimes these limits go away if we keep at it for long enough—we just need better facilities and experiments to get the answer. Often, we are pretty confident that if we could actually perform such-and-such experiment, we could resolve this-or-that mystery. Breaking new ground in modern science in this way often (but not always) comes with a big associated price tag. Next-generation supercolliders or overwhelmingly large telescopes are not cheap, but these may be required to come up with answers to some of the unsolved mysteries of the cosmos.

Sometimes our limits reflect the (relatively) extremely short time we've been doing modern science. After all, the Scientific Revolution was less than four hundred years ago, which is only 0.00000003 × the age of the universe or 0.0000001 × the age of Earth. Heck, we've only had the two pillars of modern science, general relativity and quantum mechanics, for about a century. Not only does that mean we haven't had a lot of time to figure things out, but the universe isn't set up to do a dog and pony show whenever we need data on something. The universe will take its own sweet time. Need to study a supernova in detail for your PhD thesis? Well, sit tight, odds are we will have one in our galaxy sometime in the next fifty years or so.

Sometimes the limits we encounter in trying to unlock the nature of the cosmos are cognitive. As in our own brains. Think about this: human DNA is only about 1.2 percent different from that of chimps. Chimps are smart, no question. But could you teach one calculus (not to mention general relativity and quantum mechanics)? What if our DNA were another 1.2 percent further evolved than it is? What might our brains be capable of then? The level of abstract thinking (and other types of thinking we don't even have words for) might be astounding. To be clear, I am not advocating for transhumanism. Rather, I want to flag the pure unbridled hubris involved in thinking that our brains are even capable of totally understanding the cosmos in its entirety. But that sure as heck isn't going to stop us from trying to understand what we can.

Sometimes the limits we hit are fundamental (or appear to be). There are laws of nature that we may never be able to understand, no matter how advanced our brains might become. Which means there are experiments we might never be able to perform (though I use the word "never" lightly and the word "might" with optimism). We may never be able to test what actually happens inside a black hole. We may never be able to probe (let alone interact with) other dimensions (if they exist). We may never be able to break the infinite regression of what caused the universe to be created, and what caused the cause of the universe being created, and what caused the cause of the cause of the universe being created. Turtles all the way down (we will come to the famous story of the infinite stack of turtles shortly). This is where we run smack into the boundaries of science.

For something to be considered scientific, it must, by definition, be testable. There is a tiny little loophole here: it may not need to be testable right now, but it must, at least in principle, be testable at some point in the future by some experiment that could realistically happen. If an idea or hypothesis isn't testable, that doesn't mean that it is wrong. It means it isn't testable. If it isn't testable, how do we know if it is correct? These (potentially) untestable ideas also happen to be (in my opinion) some of the most interesting ones, probably because they've been vexing humanity for millennia.

The late author and futurist Arthur C. Clarke had three adages that informed his thinking, now known as "Clarke's three laws."[7] These "laws" often come to mind when I consider the current limits of human ingenuity and experiments:

1. When a distinguished but elderly scientist states that something is possible, [they are] almost certainly right. When [they state] that something is impossible, [they are] very probably wrong.
2. The only way of discovering the limits of the possible is to venture a little way past them into the impossible.
3. Any sufficiently advanced technology is indistinguishable from magic.

Because of the unknown, and potentially unknowable, nature of the topics in this book, we will often find ourselves at the convergence of empirical inquiry (aka science), philosophy, and theology. This can be an uncomfortable space to be in, but this is also where some of the most interesting questions dwell, so let's not shy away and avoid talking about complex and loaded topics just because they are complex and loaded. The fact that they are complicated and have significance in myriad belief systems means that they really deserve to be talked about, but with a hefty dose of care and respect.

One final request before we get going: be curious. You did, after all, choose to pick up this book, so you must be at least a *little* curious. Remember all those "Why" questions you used to pepper adults with when you were a little kid? Now is a great time to rekindle that mentality and pretend that no one ever told you, "Because I said so," which is just about the most curiosity-squelching response an adult can give. The world is a lot more interesting if you are curious. Being curious is also how we make progress in science. And please tolerate "We don't know" as an answer. And by "we," I mean like all of humanity. When you get these "We don't know" answers, part of the challenge is to ask, "How could we find out?" If you crack that puzzle, you may have a Nobel Prize waiting for you down the road.

Interstition

This is a book about what we don't know and why we don't know it. But before we can talk about what we don't know, we need to think about what it means to "know" anything at all. Without taking an honest assessment of the current landscape of knowledge and understanding, we have very little hope of revealing the territory of our ignorance.

Understanding the fidelity of knowledge takes us straight to the limits of human perception and our implicit reliance on logic. But even logic may not be the arbiter of "truth" that we would like to think it is.

Thus, our journey begins by asking the humble question "What do we know?"

2

What Is Knowledge?

CONSIDER THREE THINGS YOU KNOW.

Now ask yourself, how do you actually know any of these three things?

Perhaps one of the three things on your list is your name. One could argue that your name is a good thing to know, but how do you know it? In many cases, the answer will be along the lines of your parents told you, or it is on your birth certificate. Great! Those both count as pieces of evidence. But here is the thing—either of those pieces of evidence could be faulty. For example, I'm pretty sure that in the history of humankind every child has been lied to by their parents or guardians. Mostly white lies (I hope), but lies, nonetheless. Can you be 100 percent sure that your parents told you the truth about your name? Maybe 99.99 percent sure, but that is different from 100 percent. What about your birth certificate? Are you willing to argue that no birth certificates are ever forged?

We could proceed in a similar way with most of the things you might think that you know. This is a fun and infuriating game I play with students in class. Eventually you might come up with something along the lines of "1 + 1 = 2." That *must* be true, right? How could it possibly *not* be true? Well...

In *Principia Mathematica* (written by Alfred Whitehead and Bertrand Russell),[1] it took over three hundred pages to "prove"

1 + 1 = 2, after rigorously laying out the assumptions and logic necessary to get to this crucial (and many a novice math student would argue obvious) conclusion. Even this "obvious" conclusion is based on axioms (or assumptions) that are taken to be true but cannot be proven themselves.

```
*54·43.  ⊢:.α,β ε 1.⊃:α∩β=Λ.≡.α∪β ε 2
   Dem.
      ⊢.*54·26.⊃⊢:.α=ι'x.β=ι'y.⊃:α∪β ε 2.≡.x+y.
   [*51·231]                              ≡.ι'x∩ι'y=Λ.
   [*13·12]                               ≡.α∩β=Λ      (1)
      ⊢.(1).*11·11·35.⊃
         ⊢:.(ℯx,y).α=ι'x.β=ι'y.⊃:α∪β ε 2.≡.α∩β=Λ       (2)
      ⊢.(2).*11·54.*52·1.⊃⊢.Prop
   From this proposition it will follow, when arithmetical addition has been
   defined, that 1 + 1 = 2.
```

The excerpt from *Principia Mathematica* in which Whitehead and Russell use their axioms to prove 1 + 1 = 2 on page 360 (in the second edition).

When I ask my students to tell me three things that they know (as I asked you at the beginning of this chapter), often a student will glibly volunteer that they know that they exist (often quoting Descartes, with self-satisfied certainty that *"Cogito, ergo sum"*—I think therefore I am—is definitive proof of their existence).[2] I sometimes wonder if overexposure to our own (presumed) existence has desensitized us to how profoundly nontrivial our very existence is.

If we adopt an axiom that we exist, which I admit seems about as self-evident as 1 + 1 = 2, we come to perhaps the greatest mystery of all: Why is there something instead of nothing to begin with? Not only does there appear to be something, but out of this something, life has emerged. And not just any life, but life that is capable of trying to understand the universe, why there is not nothing, and how it is that we exist at all. Back we go into the quagmire of self-reference. There is a lot to unpack here.

Now think of three things you believe.

Was that a little easier than three things you "know"?

Why do you believe these things? This *why* is important; it is the justification for your belief. Some justifications are better than others; some sources of evidence are more robust, and large amounts of evidence from different sources can build a reasonably solid case. You could, in principle, start to evaluate your sources of evidence according to how robust you think they are. This is pretty much what scientists do every day. For example, you might believe the Earth is round, which actually flies in the face of normal daily human experience. What evidence supports this belief?

The reality is, in general, we *greatly* overestimate the number of things that we *Know* (with a capital *K*). Among actual epistemologists (real people who devote their lives to thinking about and understanding what "knowledge" is), there is a lively debate about what it means to "know" something. One way to start boiling "knowledge" down is to admit that there are things that are true that we don't believe, and conversely, there are surely things that we believe that aren't true. My sense is that normal people like to think the overlap is large between what we believe and what is true, or at least that is what we often aspire to (if your aspiration is to believe things that are not true, this book may not be for you). In this belief-knowledge-truth framework, we might call *knowledge* "things that we believe that also happen to be true."

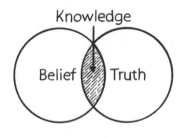

One problem that you may have picked up on is that we don't generally know which things we believe are true. If we knew they were not true, we probably wouldn't believe them (I hope). So, there is an added layer of complexity that is meant to help—this is the *justification* piece.

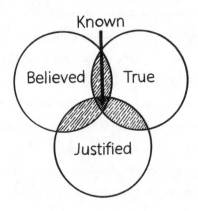

How well is your belief justified? How well does it need to be justified to be considered *knowledge*? Epistemologists even debate whether we truly "know" anything—being the mere mortals that we are, we can never really know that our axioms are actually true, or our justifications are sufficient.

The etymology of the word "know" is, at least in part, to blame for the muddied waters around the meaning of the word, which is traced back to Proto-Indo-European to a meaning more akin to "recognize" or "perceive," which are substantively different from the modern philosophical meaning. Today, we throw the words "know" and "knowledge" around with such casual imprecision, it is easy to understand how one can lose track of what we mean by the word. Some days I feel like the English language needs another word—less strong than "know" (in the modern philosophical sense), but easier in conversation than "I really strongly suspect based on justifications to date."

Imagine there's a clock tower on a route you frequently take. One afternoon you are headed to a meeting that is scheduled for 2:00 p.m. You think it is about 1:55 p.m., but you want to make sure you're on time, so you check your watch but—oh no—the battery is dead! Thank goodness, there's the trusty clock tower. Phew, you have four minutes to spare. However, unbeknownst to you, last night the clock on the tower stopped at 1:56 a.m. Fortunately for you, *it is* actually 1:56 p.m. Your belief is both apparently justified and true. But, as it happened, your justification was misguided—so did you *know* it was 1:56 p.m.? This type of scenario is at the root of an epistemological quandary known as the "Gettier problem." The takeaway message is to be aware that your justifications may not be as watertight as you might think they are. Even if your belief happens to be true, you'll be hard pressed to argue that you *know* it (at least to an epistemologist).

I will admit to being entirely inconsistent in how I think of "knowledge"; intellectually, I lean toward believing we can never truly *Know* (with a capital *K*) anything—in terms of epistemological positions, this is firmly in the realm of what is known as "global skepticism." Global skepticism gets a fair amount of flack, and for good reasons—one of which is that it is deeply impractical (although being "impractical" hardly makes something incorrect). Many people have a visceral and understandable aversion to this level of skepticism. I sympathize with these arguments. All the same, I have yet to find an argument against global skepticism that I believe is truly intellectually honest and robust against the myriad ways in which we might fool ourselves.

On the other hand, I clearly go through my daily life as though there are things that I "know" (with a lowercase *k*), because after all there is a need to be practical and proceed with our best understanding of reality. The level of cognitive dissonance this engenders ebbs and flows, and I'm OK with that—I'm human, and that comes with sometimes being inconsistent and irrational despite our best efforts.

Your Limited Senses

Before we leave the topic of what you think you might know, we should pause and consider how any information *at all* gets into your brain to begin with. We commonly think of most people as having five senses (smell, taste, sight, hearing, touch)—at least that's what was drilled into my head in elementary school.

Certainly, there are lots of people who don't have one or more of these senses for a variety of reasons, and the ways in which their brains can literally rewire to perceive the world in new ways is amazing. On a modest scale, if you happen to wear glasses and you've ever had a new optics prescription, you may have experienced the world being oddly distorted, even causing headaches or issues with depth perception. Eventually, your brain adapts and everything is "normal" again. The point is that your brain is doing the work of constructing your perceived reality, which may or may not be an accurate reconstruction of actual reality.

We know of other animals that have evolved with the ability to detect all sorts of cool things we humans cannot. Take for example the mantis shrimp; the mantis shrimp not only detects polarization of light, but while we simple humans typically have what is called "trichromatic vision" (three types of color receptors in our eyes, roughly red, green, and blue), some species of mantis shrimp have sixteen![3] Curiously, some people are born with a fourth type of color receptor, making them "tetrachromats," which seems like a superpower.[4]

Are there other things that our five senses have not evolved to detect? How might we perceive the world differently if we could? The fact is, we simply do not have direct access to reality with our senses. This is acutely important to keep in mind when contemplating how much of the universe we understand.

My goal here is to entice you to start questioning your own knowledge, and why you believe the things that you do. As the late physicist and famed Nobel laureate Richard Feynman is quoted as saying, "The first principle is that you must not fool yourself, and you are the easiest person to fool."[5] Here is a question to ask yourself: What evidence would cause you to change one of your beliefs? For example, what would be required to convince you that your birthday is not when you think it is? Or that the Earth is flat? Or that you are a computer simulation? If you could, in principle, change your belief based on new evidence, then your belief is said to be "defeasible." This is a great word, and at the core of science. I challenge you to try to work this word into a conversation at some point today, just for fun. To make progress in science, we must be willing to change what we believe to be true about the universe based on new evidence. And we are always getting new evidence. The general goal is to get those three circles in the Venn diagram of "believed," "true," and "justified" to overlap to the greatest extent possible.

The problem is that changing what you believe is hard. In fact, our brains are wired with a tendency to both seek out and interpret information in ways that support what we already think (I mean, just look at social media). This is stunningly shortsighted of our brains and can leave us deeply misled. We also tend to shy away from experiencing *cognitive dissonance*, which is that stressful feeling when you are trying to hold inconsistent ideas in your head at the same time. To solve some of the greatest mysteries in the universe, we must be willing to push ourselves out of our comfort zone and try to make sense of things that don't appear to make sense.

What We Take for Granted (Axioms and Cats)

As much as we might sometimes hate to admit it, we must make a host of assumptions in daily life. Yes, it is much better to have actual facts that we *know*, but we must also acknowledge that things we truly know with certainty are rare. When it comes to formal logic, we call these assumptions "axioms." An assumption doesn't have to actually be true to be an axiom, it just has to be something you take as given for the sake of an argument (or logical structure). For example, you could take $1 + 1 = 2$ as an axiom, which for most normal people seems self-evident (although as you read earlier, Russell and Whitehead might beg to differ). You could also take $1 + 1 = 0.5$ as an axiom. That seems less self-evident. But given the assumption that $1 + 1 = 0.5$, you *could* hypothetically try to start building a logical structure based on that and see what happens.

When it comes to trying to understand the universe, we tend to think of axioms in two ways:

1. The standard axiom: things/ideas/concepts that we generally believe are true. If we assume these things/ideas/concepts are true, there are implications for how we understand theories and interpret data.

2. The "What if?" axiom: things/ideas/concepts that we don't necessarily think are true, but what if they were? This type of thinking, of questioning fundamental assumptions,

is often what leads to the biggest breakthroughs in science. It also requires creative thinking and gaming out scenarios.

We rely on axioms in our daily life just to make sense of the world. When my children were infants, I was fascinated watching them interact with the world around them and slowly start building up the expansive set of things we just take for granted as *the way things are*. I suspect that you might also take it as given that gravity pulls things down, that you are alive, or that time always flows at the same rate. Really, any fundamental things you might take as self-evident get built into your worldview. How would your worldview have to be different if a particular axiom you held were untrue? Whatever the alter-axiom is could be a fun premise for a speculative fiction novel.

Some of the most famous axioms are often referred to as "Aristotle's laws of thought" (although to give appropriate credit, similar ideas are found in Indian Logic back to the sixth century BCE). I think you will agree that these axioms seem pretty self-evident, especially when it comes to cats:[6]

Name	Axiom	Example	Cats
Identity	A thing is equal to itself.	$A = A$	A cat is a cat.
Excluded Middle	A thing must have some characteristic or not have that characteristic.	$A = B$ or $A \neq B$	The cat is alive or the cat is not alive.
Non-contradiction	A thing cannot both have a characteristic and not have that characteristic.	A cannot both $= B$ and $\neq B$.	The cat cannot be both alive and dead.

These "laws of thought" seem solid, right? Because these axioms seem so exceedingly self-evident, it is hard to imagine how they could possibly be untrue. But then this physics theory called "quantum mechanics" makes a spectacular entrance.

If you haven't encountered quantum mechanics until this point in your life, you stand a better chance of being a nice, normal, well-adjusted person. As the late Richard Feynman is often credited with saying, "If you think you understand quantum mechanics, you don't understand quantum mechanics."[7]

Why does quantum mechanics cause Nobel Prize–winning physicists to question reality? Well, the math is complex, sure, but that is not out of the ordinary in high-level physics. The real issue with quantum mechanics is that it calls into question those extremely obvious and self-evident laws of thought. For example, in the most well-known example of a cat making a cameo in physics, Erwin Schrödinger proposed a thought experiment (no actual cats were harmed, only hypothetical cats) in which, according to one interpretation of quantum mechanics, a cat is both alive and dead at the same time. The poor cat is suspended in this state of superposition until the box it's in is opened and the quantum mechanical description of the cat is forced into alignment with Aristotle's laws of thought.

There are plenty of other cat-free examples of axioms that might seem obvious in one frame of reference but get cruelly smashed when we change how we think about things. Take Euclidean geometry for example. Remember the axiom that parallel lines never intersect? Draw two parallel lines on a piece of scrap paper and test it. Despite the fact that your lines are probably *not exactly* parallel because you are human, your test probably supported this axiom. So, this axiom works great. As long as you are drawing on a flat surface. If you happen to have something spherical at hand, go try the experiment. How'd that turn out for you? There are a whole set of axioms that work perfectly well in flat space (also known as "Euclidean space") but fail if we try them in other situations. The lesson embedded here: just because an axiom seems to hold true on your kitchen table, doesn't mean it holds true beyond our experience or ability to test it.

The point is that even axioms that seem so blatantly self-evident as the laws of thought are not sacrosanct. And that should make you question your grip on reality. Axioms that may appear to hold up in your tiny little corner of existence may well not be universal truths. The sooner you internalize that, the better.

Still, here we are trying to do our best to understand reality. Instead of just throwing up our hands and giving up, we've leaned into empirical inquiry to test our assumptions and ideas when we can.

Empirical Inquiry on a Deserted Island

At this point, you may well be thinking, "Then how do we know anything at all?" Good question. Maybe we don't. But trying to figure things out is more fun than giving up. Also, it helps us stay alive (we think, but that is an unprovable axiom).

Imagine a scene out of a bad movie in which you're on a plane that has crashed into a remote jungle and you are trying to survive. How do you figure out what you can eat without poisoning yourself?

Here is one option: figure out what other animals are eating and what they are avoiding. That bright yellow berry no animal has touched? Maybe not the best idea to eat. Those lovely, sumptuous fig-like things the unidentifiable reptile is eating? Could work.

Are you sure if you pop it in your mouth you won't die? Better test it first. There is actually a whole long process called the "universal edibility test." It goes roughly like this:

a.   Smell it. If it smells really bad, don't eat. If not, continue.

b.   Put some on your skin.

c.   Did you get a rash, go numb, or start burning? Don't eat. If not, continue.

d.   Prepare a small portion and touch it to your lips and wait. See (c).

e.   Put a small amount in your mouth and wait. See (c).

f.   Swallow a teeny tiny bite and wait (like *hours*). Are you still alive? Continue with cautiously eating.*

---

* If you plan to get lost in the jungle, please find actual survival advice from experts and not an astrophysicist.

What if one of those tests failed, but you're still alive, and now even more hungry than you were? Guess you better reassess whether that unidentified reptile had the best choice of culinary habits to mimic. Oh, look over there—that thing looks like a deer! At least it is a mammal. Let's go watch what it eats (and what it avoids). But even that isn't foolproof—for example, the aptly named "deadly nightshade" (or belladonna) can be fatal to humans, but other mammals, like cows and rabbits, don't seem bothered by it. Or poison ivy, which goats can munch on unharmed with wild abandon. On the other hand, there is chocolate. If you find a chocolate bar on the island, that is a real treasure and may have you questioning your "deserted island" situation. But many animals (including dogs, cats, and bears) can't metabolize a chemical in chocolate called "theobromine," which is toxic to them.

The morals of the deserted island lesson are (1) be very careful about the assumptions you make in the process of empirical inquiry, and (2) also maybe don't get stuck on a deserted island.

The universal edibility test is a great example of the scientific method. The thing about the scientific method that can get annoying is that you never actually *Prove* anything (with a capital *P*). All you can do is keep throwing harder tests at any given explanation to see if it passes, which brings us to "the loop." Test the theory. If the theory passes the test, come up with an independent or more challenging test. Repeat. The more times you go around the loop, the more confident you are in your explanation. Been around the loop a single time with an easy test to pass? Maybe don't eat the fruit-like thing you found just yet.

We're getting knee deep in some science vocab here, and it's important that you understand what scientists mean when they use certain words (it may not surprise you to know that what scientists mean is sometimes different than what you might encounter in the normal world of people who don't care why gravity works).

A hypothesis, a theory, and a law walk into a club looking for the perfect explanation:
Hypothesis says, "I think you're really smart."
Theory says, "Let's go on a few dates."
Law says, "I'm ready to commit for life."

"Hypothesis" is a fancy word for "guess." Even better if it is an educated guess or based on some insight or intuition. If all your hypotheses turn out to be correct, you are probably not thinking creatively enough. If solutions to unsolved mysteries were obvious, we would have figured out the answers by now. So, don't shy away from crazy ideas. Crazy ideas sometimes turn out to be correct. But your crazy idea does need to be, in principle, testable. Is your idea testable? Congratulations, you may proceed with doing science.

"Theory" does not mean the same thing to scientists as it usually does in a crime show on TV. If you're chatting with your nonscientist friend and they say they "have a theory" about something, probably what they mean is they have an educated guess. We have a ten-letter word for "guess" that starts with *H*. When a scientist uses the word "theory" in a scientific context (and not at a party, where we sometimes try to blend in and speak like normal people), what they mean is an "explanation" that appears to fit the facts and data *so far*. A good theory needs to be able to make predictions that can be tested, and you never know—the next big test of the theory might break it. The harder the test is to pass, the better.

Once we've tested a theory over and over and over (going around and around the loop), and the tests have gotten harder and harder, we might finally call it a "law," which is code for "This theory has passed every test with flying colors, and we are pretty much running out of ideas to test it." In this context, a law is as close to something being proven as science gets. However, be warned, even in science, there are a lot of things and concepts that are mislabeled, because what fun would it be if there weren't exceptions to the rule? For example, the theory of general relativity is pretty much as close to a law as anything we have, but it has "theory" right there in the name, just to confuse unwary students. Likewise, Newton's law of gravity is not only not a law, it isn't actually correct—the law of gravity does a passable job here on Earth, but ultimately it is just an approximation that only works under limited conditions (not to throw shade at Newton or cause you to distrust your high school science teacher). But don't feel too bad for

Newton—he did get a "law" named after him, and that is a big accolade in science.[8]

Can We at Least Disprove Things?

If we can never actually *Prove* anything (with a capital *P*), can we at least disprove things? Then we could effectively prove things by disproving all the other options. As Arthur Conan Doyle famously wrote, "Once you eliminate the impossible, whatever remains, no matter how improbable, must be the truth." One clear problem with this approach is that it relies on you knowing *every other* possibility. It also depends on your ability to disprove these other options.

Disproving things gets complicated, and how we address this depends on whether we are talking about being practical or being sticklers for strict logic. Let's go back to the deserted island scenario for an example. You have your hypothesis that because some mammal-like creature appeared to be eating that fruit-like thing, it might be safe for you to eat. So, you rub some on your skin and break out in a nasty rash. Does that disprove your theory that you can eat it? For practical purposes, probably yes. But we can also hypothesize other explanations for what might have caused the rash besides the fruit-like thing being poisonous to you. For example, (maybe) earlier in the day, unbeknownst to you, another animal with poisonous saliva spit on the fruit-like thing, in which case, all you need to do is wash the fruit-like thing off before eating it. We can come up with increasingly far-fetched explanations, but if you are trying not to die in the jungle, maybe don't venture too far away from the practical.

From a purely philosophical standpoint, if you cannot prove anything to be true, that means you cannot prove any of your premises to be true, or any of your experimental steps to be unflawed (because these things also fall under the very large umbrella category of "anything"). Ultimately, this line of logic means that you cannot 100 percent prove anything to be false.

Here we run smack into a principle known as "Occam's razor," which is often turned to as an arbiter of the "best" explanation. But be wary, valiant reader: just because something is the "best" explanation, does not mean it is *correct*. If you haven't encountered Occam's razor, it goes like this: for any given set of data or observations, there are countless possible explanations. But the explanation that requires the smallest number of ad hoc assumptions is *most likely* to be correct. Note: I did not say that the particular explanation at hand *is* correct, only that in a statistical sense, the greater number of crazy assumptions we have to make, the *less likely* that explanation is to be correct.

Let's try an actual scenario:

Say a professor came to class from their office. The simplest path they might have taken could be to have walked directly on the shortest possible route between their office and class. On most days this might be what they do. But perhaps today, they needed to stop by a colleague's office first to drop something off. Or maybe they desperately needed coffee before trying to teach the class something (does the professor have coffee with them?). Or maybe, they went by their colleague's office, they both went to get coffee, then decided they had time for a nice stroll around the campus pond before the professor came to class. None of these paths are totally outlandish. But if you had to place a bet, without any additional data, which one would you pick? If your professor is carrying a coffee cup with a store logo, that observation could count as "data" that needs to be accounted for in your explanation. Or if the professor walks in with their shoes soaking wet on an otherwise dry day, you might try to build that into your explanation, making the pond route ever more likely.

We could add in some much crazier paths. For example, the professor left their office very early this morning, flew to a different city, gave a talk, and flew back just in time for class. Is it possible? Sure. Likely? Probably not. Unless the professor walks into class with a suitcase and exclaims, "I just got back from DC where I gave a talk!," for example. But the simplest explanation that fits the data you have in hand is generally the most likely to be correct according to Occam and his razor.

Ultimately, when it comes to proving or disproving things in science, you will find researchers saying a lot of things like:

"The results are consistent with the theory of . . ."

This phrasing is code for "The data might support the theory, but we are not willing to go so far as to use the word 'prove.'"

Or, if you are not averse to double negatives, a time-worn favorite:

"The results are not inconsistent with the theory of . . ."

This phrasing is code for "The data don't rule out *or* support that theory, but we want to sound more scientific than just saying it that way."

However, sometimes crazy things turn out to be true, and we have to be open to the next experiment throwing a wrench in our theory. Have you ever heard of the Mpemba effect? Probably not. Under very special circumstances, hot water can freeze faster than cold water. Does that defy your intuition and lived experience? Good. Let that be a lesson.

Positivism Run Amok

If you haven't encountered positivism before, let's start by just nipping an understandable misconception in the bud; positivism has nothing to do with being a happy and optimistic person (although I suppose one can be both positive and a positivist, but I know a lot of cranky positivists). The term "positivism" is derived from the French *positif* (which, in turn, comes from the Latin *positivus*), roughly meaning "to put or place," as in "*posit*ion." In this case, being "positive" takes the valence of being "certain." In terms of philosophy, positivism refers to the conviction that the only meaningful measure of reality is obtained from logic, reason, and experiment. Given this definition of positivism, I hope you can see that it has an important role in science and the scientific method.

I'm pretty much on board with much of positivism (which has many variants, not unlike the range of Protestant denominations, which agree on some overarching things and differ on a host of more nuanced issues). My own belief in the importance of empirical inquiry is not surprising, given that I'm a scientist. However, I have my limit. There is an extreme version of positivism that contends that *no meaningful reality exists* outside of logic, reason, and experiment. To my ear, this version of positivism is dripping with unjustified confidence in the narrow range of things we humans have evolved to do, sense, and think about. Denying any reality other than that which we can perceive and test strikes me as myopic and a path to dead ends. If we aren't at least a tiny bit open to things outside our currently empirically testable reality being true, we are cutting ourselves off from all sorts of possibilities—possibilities that may one day even be testable. Cue the second of the late Arthur C. Clarke's three laws, "The only way of discovering the limits of the possible is to venture a little way past them into the impossible."

Because of my own philosophical position, I am willing to entertain ideas that are not currently testable (and may never be testable) as part of reality. But considering something and believing something are not one and the same.

Logic, Fallacies, and Ice Cream

Trying to understand the universe also depends on our ability to apply logic, which is one of the most important assets among the meager tools humans have. The problem is that our human brains also fall prey to tricks and shortcuts that can lead us astray. I won't go into an exhaustive list of logical fallacies here (you're welcome), but just highlight a few that seem to be common when discussing science and the universe.

Correlation and causation. As the saying goes, everyone who confuses correlation and causation dies. This fallacy also goes by the much fancier Latin term *Cum hoc, ergo propter hoc*, if you want to impress people at parties (make sure you say it with a pretentious accent). The Latin translates into "with this, therefore because of this." We get tricked by this fallacy because lots of things appear to be related (we call this "correlated") because they *are* causally connected. For example, there is a causal relationship between how much cayenne pepper I accidentally spill in a recipe and how spicy the resulting meal is. We could make a plot of these variables, and there would be a nice tidy correlation that is, in fact, causal. However, the inverse is not true: a nice tidy correlation does not require causation.

For example, there is a correlation between ice cream sales and violent crime rates; the hot summer months see the most action in each.[9] Because ice cream sales and violent crime rates are both causally connected to the season, they *are also correlated with each other*, and this is where our brains can get tricked if we are not careful. Because sometimes correlations do, in fact, imply causation, it is tempting to take a shortcut and assume they always do. In the ice cream and crime example, you might (wrongly) infer that violent crimes cause people to want to eat ice cream, or alternately, that eating ice cream causes people to commit violent crimes, and that would just plain suck.

After this. Now that you're getting your Latin down, here is a related fallacy: *Post hoc, ergo propter hoc*, or "After this, therefore because of this." The difference between this fallacy and *Cum hoc, ergo propter hoc* is that here we aren't talking about correlations but rather discrete causal events. For example, there used to be a cupcake bakery where I live that made special lemon cupcakes that were said to induce labor in women in the late stages of pregnancy. The bakery advertised great numbers of women who had given birth within forty-eight hours of eating one of these magical lemon cupcakes—and I am one of them. What else could have possibly caused all of us to go into labor? Shockingly, pregnant women who are past their due date tend to give birth pretty soon. In my case, I was already almost two weeks overdue with this baby when I was gifted the lemon cupcakes. Still, lemon-drop cupcakes are a thing for pregnant women (go ahead, search the internet, it won't disappoint). Maybe it is just a thinly veiled excuse to eat delicious cupcakes, and if you are pregnant and two weeks past your due date, as far as I'm concerned, you've earned those cupcakes.

We are exceedingly easy to dupe with this fallacy because we really like tidy causal relationships, especially when things might seem outside of our control. The bottom line is that just because X happened after Y doesn't mean that X caused Y.

False dichotomy. You're with us or you're against us. Sound familiar? Parsing the world into separate little buckets of opposites sure does help to simplify things. Too bad it doesn't always work.

There are two different ways you might visualize the options besides "with us" and "against us." It could be that "with" and "against" are clearly defined categories, but there are positions that don't fall in either. It could also be that "with" and "against" are just the opposite extremes on a continuum, and people can fall anywhere in the middle (like being 90 percent with us or 60 percent against us).

It could also be that we just aren't aware of other categories or possibilities. Imagine that you are completely color-blind and only see shades of gray. In this case you would be able to put the gray level of everything you see on a scale between white and black. But then someone told you about color! You've never seen color, so this is an abstract concept, but suddenly your simple grayscale system needs to have an entirely new dimension to it. The lesson is that *assuming* you know all the options based on your experiences is not solid ground.

To be clear, there is nothing special about "dichotomy" (meaning two options), as opposed to "trichotomy" (three), "quadchotomy" (four), and so on. It really doesn't matter how many bins we create with labels to put things in, most things in the universe fall on a continuum. A classic example is the rainbow. I grew up learning the rainbow song lyrics "Red, orange, yellow, green, blue, purple / Those are the colors of the rainbow." Maybe that is true for elementary school color theory, but it is also fundamentally wrong—the rainbow could be subdivided into a virtually infinite number of colors (down to the quantum limit) and extends far beyond what the human eye can detect (admittedly, this extended rainbow makes for a terrible song to teach elementary schoolkids). Because there are six categories of color in the elementary school case (or seven if you learned your colors as

"Roy G. Biv"), you could call it a "sexchotomy," but I don't see that term ever coming into fashion.

We firmly encounter the physics version of the false dichotomy fallacy when we think about whether light is a particle *or* a wave; it turns out that the nature of light isn't so straightforward. In the case of light, in hindsight we can see that our preconstructed dichotomy broke down. To this end, when we are thinking about possible resolutions to the mysteries in this book, we need to have "or something we haven't thought of yet" as an option.

Argument from ignorance, also known as the " 'God of the gaps' fallacy." Simplified, this fallacy goes along the lines of "If I don't understand something, it must be due to a supernatural being." So baldly stated, it seems obvious that a vast category of possible options is excluded, which might be called "things that are not supernatural and have explanations that I don't understand." Or more generally, apropos of this book, "things that humans don't understand (yet) but that are not supernatural." Note: this isn't to say (and I am not saying) that supernatural things don't exist. But rather, assuming you are clever enough to understand everything in the universe that isn't supernatural is . . . well . . . stupid.[10] Can you think of examples of this fallacy that you've encountered? If not, please go search on "Bill O'Reilly and tides."

In fact, there is a danger in using this argument to support the existence of the supernatural, as articulated by the German anti-Nazi theologian Dietrich Bonhoeffer:

> How wrong it is to use God as a stopgap for the incomplete-
> ness of our knowledge. If in fact the frontiers of knowledge
> are being pushed further and further back (and that is bound

to be the case), then God is being pushed back with them, and is therefore continually in retreat. We are to find God in what we know, not in what we don't know.

The take-home point: just because you don't understand something doesn't mean a supernatural explanation is required. The corollary is also true: just because you think you understand something does not prove divine intervention wasn't involved. So be careful. If it were easy to prove or disprove the supernatural, we would have done so a long time ago.

Equivocation. Some words or terms have more than one meaning, and this allows them to be misused in a way that can make arguments misleading. For example, in science we run into this a lot with the word "theory." As discussed at the beginning of this chapter, when scientists use this word in a scientific context, it means "a hypothesis that has been tested quite a bit and it has held up." But in the colloquial sense, "theory" is used to mean something more like a guess. This leads to situations like people saying, "You shouldn't believe in the theory of evolution because it is only a theory."

Genetic fallacies. A large set of fallacies fall under an umbrella that impugns their origins. Indeed, called "genetic fallacies"—from the Greek word *genesis*, for "origins"—they are about mistaking the credibility of a source for a measure of an idea's truth. One of these, the fallacist's fallacy, assumes that a conclusion is false because the argument for it isn't valid. Another, the appeal to authority, says that something's truth value depends on who makes the claim. And then the ad hominem, which is not an attack on an idea, but simply on the person who made it.

Aside from the practical value of finding these fallacies in daily life, it is essential that we try to keep these fallacies at bay as we embark on thinking about great mysteries in the universe. When our brains have very little actual information to use, they are extra good at jumping to conclusions, seeing patterns, and drawing connections that may not exist.

Every one of the coming chapters requires us to think carefully about what it means for us to "know" anything. Empirical inquiry, math, and logic can take us to the outer reaches of modern science, but some of the most vexing questions lie beyond the reach of these tools. As we embark on a journey through some of the bewildering unsolved mysteries in the universe, we will frequently refer to this chapter and if, or how, we know anything at all.

Interstition

Now that we have a little epistemology under our belts—and what it means to *know* anything—we can begin to approach the *unknown*. To be sure, answers to the great mysteries of the universe are still in the category of "unsolved" for different reasons: some of these mysteries may ultimately be beyond the ability of our human brains to understand. Some may be solvable in the not-so-distant future, merely requiring more time, better data, and clever analysis. Still others may require novel breakthroughs in math or physics outside of our current scope of knowledge. The following three chapters will focus on a trio of questions, one from each of these categories: one we may never know the answer to, one we may just need more time and data to solve, and one that seems to require new breakthroughs in physics to make progress.

I can think of no better place to start than the beginning.

3
What Caused
the Big Bang?

"MOMMY, I HAVE A QUESTION."

Buckle your seat belt, here we go. This may sound like a simple statement coming from my then four-year-old daughter, but now after three children I am acutely aware that the questions that can come from young children can range from apparently nonsensical to unanswerable.

The "Mommy, I have a question" megaquestions had become commonplace, and my daughter's favorite time and place to ask these zingers seemed to be on our long drive into town. It is hard to *really* explain something at an appropriate level for a four-year-old regardless, and it doesn't help when you don't want to crash the car.

The week before, my daughter followed up one of these "I have a question" declarations with "Where did the first mommy come from?" Questions like this send my internal dialogue into a frenzy—how do I answer this fairly, honestly, and at the right level for a tiny human who doesn't have the education, vocabulary, or abstract thinking skills yet to tie her own shoes?

On this particular morning it was raining as we drove along the country road. My daughter had been sitting quietly for some time, watching the raindrops slide across the window.

"Where does water come from?"

Easy! I have a PhD—I can totally explain the water cycle to a preschooler, right? Ah, hubris. I commence to try to explain evaporation, condensation, and precipitation *without* actually using any of those multisyllabic words, and I was feeling pretty proud of myself. I was low-key, thinking she would get bored with my explanation and abruptly change subjects. Non sequiturs and kids go hand in hand. But in the rearview mirror, I can see my daughter is not looking happy about my explanation (I used to call it her "crumple face"). In fact, she appears to be getting increasingly frustrated as I continue. I think maybe I'm using terms that are too complex or ideas that are too abstract, and probably I am—because really, trying to explain the water cycle to a four-year-old while driving is not trivial. I try out some simpler words and concepts—like rain.

"Sweetie, you know how sometimes it rains?"

"Noooo! That's not what I mean!"

In the reality of the moment, it's clear that I've misread my daughter's query, and I'm trying to figure out what she really *meant* to ask. I would dearly like to avoid a tantrum, which I can see simmering just under the surface. Her frustration is mounting with my lack of understanding, which is totally fair. It's never good to start the day with screaming and crying (which goes for me, too).

Now, with her annoyed tone, and a level of petulance that only toddlers are capable of using and getting away with, "No, Mommy, where does the water *come from*?"

Um, OK, maybe this question isn't as easy as I thought.

"Sweetie," which I say in my most calm, most patient mommy voice, "do you mean why do we have water on Earth to begin with?"

"Yes! Where does the water come from?"

It turns out that she had expressed exactly what she meant to begin with. "Where does water come from?" In fact, I had just assumed that she wanted to know about rain because it was raining. She didn't ask about rain, she asked about water.

Finally, I think I've grasped the core of her question. Well, I'm an astronomer, so I've got this one covered.

"That is a great question! There are these huge dirty snowballs in space called 'comets,' and when they crashed into Earth a *loooong*, long time ago they brought us their snow and ice, which is where a lot of our water comes from." OK, so that explanation is not 100 percent precise—for example, there is no actual "snow" in space, and other types of planetesimals brought water too (the precise fraction of water brought to Earth from different sources is debated), but this was the best I could come up with on the fly, so cut me some slack. Sometimes we have to sacrifice being precisely correct for the sake of moving the understanding of students (or a toddler) forward, and hope these "analogies" don't come back to bite us later.

My daughter still has that crumple face though. Now I am thinking she's going to worry about comets crashing into Earth. At this point my internal dialogue is prepping me to answer a question about why we don't have comets crashing into Earth all the time any-more, and I'm really not prepared to try to explain probabilities and statistics.

Now I get her super-I'm-an-angry-toddler tone, "But where does the water *come from*?"

My poor morning mommy brain is clearly far behind hers—she literally means what she is asking—she wants to know where water *comes from*—as in before it came to Earth, before the comets and asteroids had it, before water even existed as water. That is a big ques-tion. I clearly had not had enough caffeine yet, but I finally caught on—we were about to dive headfirst into an infinite regression of sci-entific knowledge and understanding, the chain of cause and effect back to the "beginning."

By the time we made it to my daughter's daycare, we had worked back through the formation of molecules in giant clouds in space, nucleosynthesis in the cores of stars, and eventually all the way back to the grandmother of them all—the beginning of the universe. (Again, not actually using any of those words, and sweeping A LOT of stuff under the rug. It turns out to be hard explaining nucleosynthesis to someone who doesn't know what an atom is.) At that point we hit the wall (figuratively speaking at least, fortunately by now I had parked the car).[1]

"Well, sweetie, nobody *really* knows how the universe was created. Lots of people have lots of different ideas, though." I was certainly tempted to use the word "hypotheses" here instead of "ideas," but I held back. I was already struggling to explain the topic at hand, and I did not want to get into even deeper trouble by using a word that would cause more problems.

"Like what ideas?"

Hmmm. I am seriously not up for trying to explain quantum fluctuations to a four-year-old. With a nod to the great minds that have preceded this discussion, and the volumes of text devoted to the topic of how the universe came to be, we still have a four-year-old in the back of the car who wants to know where water (and now also the universe) comes from.

Turtles All the Way Down

This anecdote about my daughter gets straight to the concept of "infinite regression"; for each element in the chain, there is a preceding element that caused it. Eventually we might get to "What caused the Big Bang?" Let's say, hypothetically, I could give you an answer to that. Your next question might then be along the lines of "What caused what caused the Big Bang?," and thus we find ourselves in a chain of infinite regress. Different stories have incapsulated the idea of infinite regression in the form of "turtles all the way down," which references the Earth being on the back of a turtle, which is on the back of another turtle, and on it goes. All the way down. I would argue that modern cosmology, with all our amazing data, complex math, and high-powered telescopes, is fundamentally not in a very different philosophical state.

Not only does the topic of the Big Bang get us mired in infinite regression, but causality itself comes into question. Given our normal human experiences in this mundane place and time in the universe, it is easy to take causality for granted—causality is deeply engrained in our existence and has been reinforced over countless millennia of evolution. Every event seems to have a cause (or set of causes that combine to produce an outcome). My favorite coffee mug broke *because* it fell. My plant died *because* I didn't water it. Of course, for each of these (true) examples from my daily life, the actual causes are much finer grained, having to do with microscopic interactions of forces happening beyond human senses affecting atoms and molecules, but the point is that macroscopic causality is everywhere and every-when around us. One of the issues when discussing the Big Bang is the nature of time (in a chapter coming soon), and whether time itself had a beginning. If time itself began with the Big Bang, talking about a notion of causality *outside of time* is problematic. To be fair, exactly what causality is in the abstract is a subtle notion that philosophers have been wrestling with for centuries, and the true underlying nature of what—exactly—causality is may be at the center of the Gordian knot we find ourselves in.

To put a finer point on this, here are issues that have kept many a physicist, philosopher, and theologian up at night:

First, consider the conjecture that time is infinite toward the past, that is to say, time had no beginning. In this case, what caused the universe to be created when it was? Why was that particular point in time (among an infinite number of points in time) special? If there had been an infinite amount of time preceding the birth of the universe, why had the universe not previously come into being? In fact, if there was an infinite amount of time in the past, why has everything not already run down à la the second law of thermodynamics?

In the thirteenth century, Thomas Aquinas tapped into arguments in this vein to conclude that God exists, time had a beginning, and that God exists outside of time.[2]

For the sake of full disclosure, I have a huge intellectual crush on Aquinas; I think he would have loved modern cosmology (and perhaps also been bemused that, centuries later, our philosophical understanding is basically the same as the view he endorsed). But to be sure, Aquinas was not the first to consider these issues. As just one example, nearly a millennium before Aquinas, St. Augustine was a little salty on this topic. There is a famous anecdote about Augustine being asked what God was doing before he created the universe, to which in the apocryphal account he supposedly replied, "He was preparing hell for those who pry into mysteries."[3] So, I guess I'm in trouble then.

Next, consider the conjecture that time had a beginning. In this case, we quickly run into issues not only with causality (how can there be causes if there is no time?), but also with our use of language—can we even describe things "happening" outside of time? For example, one might be tempted to ask, "What caused time to begin?," but in the absence of time, it isn't clear that this question even makes sense, and we end up in a bit of a logical and linguistic tangle. The primary way I have tried to personally make sense of this is to think about dimensionality more broadly; yes, we currently perceive three dimensions of space and one of time, but I need to envision "extra" metadimensions outside of these, in which our perceived reality exists. These metadimensions may have quasi-spatial or quasi-temporal properties, a combination of these, or some other flavor of dimension that we don't ever perceive or have language for.

To be clear, I have no cause to believe such metadimensions exist, but without them my intuition hits a wall. In fact, I have a niggling sense that these questions have answers that live in a landscape that is just beyond the ability of humans to grasp, which makes me acutely aware of our own intellectual and sensory limitations, and that paltry 1.2 percent we are different from chimpanzees. But as long as I bear in mind that this line of thought might be completely deluding myself, I've decided I feel OK using it to help me imagine possibilities. If we allow ourselves to pursue this as a thought experiment, we can then give ourselves permission to imagine a quasi-temporal dimension outside of the universe and outside of the conventional dimension of time. Then we can make up new language to describe positions in this metadimensional landscape, and we can imagine causes that occur outside of time in the universe. This enables us to formulate questions like "What caused time to begin 'quasi-when' it did?" and "What existed 'quasi-before' time?" I know, the phrasing isn't elegant, but you try making words to describe events in dimensions we don't perceive and that may or may not exist. I suppose we could abbreviate this notion of quasi-time outside of conventional time as "q-time," but I am not sure that is any better.

Before we even start this discussion of the Big Bang, I want to acknowledge that this is thorny territory; there is no getting around the fact that the topic of the beginning of the universe goes straight to the overlap of science, theology, and philosophy; it is thin on empirical data, and ripe for debate and disagreement. My experience has been that topics like this are often avoided in conversation because they can lead to situations like nasty debates at family reunions, but I believe that questions like this can be the most rewarding to explore.

There is a prevailing popular assumption that scientists are not religious. I can assure you that this assumption is wrong—I know scientists who have beliefs across the spectrum. Heck, the Vatican even has an astronomical observatory in Arizona with Jesuit astronomers. I personally take it as self-evident that there are things that are true about the cosmos that we cannot test or prove (at least not yet). Just because something is outside the realm of empirical inquiry doesn't make it impossible, it simply makes it untestable. Many people's beliefs about the universe are fundamentally anchored in the foundational axioms they hold (some of which they may not even be aware of) and what they consider to be valid justification. This goes back to the early chapter on epistemology. If you skipped that, I would be grateful if you went back and at least skimmed it because it will keep coming up throughout the book.

I went to graduate school in astrophysics alongside a fellow student whose fundamental axiom was that a god created the universe in its current form less than 10,000 years ago. He contended that the evidence we have for the universe being much older (by more than 13 billion years) was essentially put in place by a higher power to mimic an old universe and fool the nonbelievers. The overwhelming majority of astronomers I know cannot abide young-Earth creationism because it does not stand up to empirical inquiry. Test after test indicates that the universe is almost 14 billion years old. However, if we are being intellectually honest, we also must acknowledge that we cannot actually disprove that the entirety of the universe is a contrived set-up job by a higher power to fake us out. That is the thing about fundamental axioms—they themselves cannot be proved or disproved. Given the outsized impact of axioms in shaping our beliefs and worldviews, it is essential that we come to terms with our own axioms, and why we hold them.

For this reason, as we talk about the Big Bang, I want to be transparent about why we think we know what we think we know and what the limits of our current knowledge are. Ultimately, we will find ourselves in the borderlands of science where the hypotheses we devise are currently untestable.

What's in a Name?

There are a host of misconceptions out there when it comes to the Big Bang theory, some of which are caused by the name itself. For starters, the word "theory" that colloquial language suggests is some random idea that somebody's uncle had. Even worse than the word "theory" is the phrase "Big Bang," which erroneously gives people the impression that this event was like a big explosion in space somewhere. We need to clear this up; it wasn't an explosion, but rather a rapid expansion of space itself. Moreover, it didn't happen at a specific point in space, but rather happened to all of space at the same time. In other words, wherever you are at this very moment, the event known as the "Big Bang" happened right there 13.7 billion years ago. It also happened where I am sitting here writing this paragraph, and in the most distant galaxies we can observe. The Big Bang was an event *of* space, not *in* space.

What astronomers mean with the term "Big Bang" is a brief period of unfathomably dramatic inflation of space itself that happened in the very early universe. We can put some numbers on what astronomers refer to with this dramatic inflation; specifically, linear scales in the universe expanded by a factor of 10^{26} during this inflation. In other words, if the size of something before inflation had been 1 mm, during inflation it would have inflated to the size of 10 million light-years.

What matters is not just the change in size scale during inflation, but the timescale over which this "cosmic inflation" happened. Current evidence suggests that this dramatic expansion happened at a time of about 10^{-36} seconds after the universe came into existence and lasted for less than 10^{-32} seconds.[4]

I don't even have a bad analogy to help you intuit how short that timescale is in a human frame of reference.[5] You may have noted that this expansion of points in space happened way faster than the speed of light. Lest you worry that we have abandoned a sacred law in physics, this rate of expansion is allowed to be faster than the speed of light because *no information was conveyed*. Things in the universe can move (and have moved) apart from each other faster than the speed of light, and this is OK. What they cannot do is have causal contact with each other (i.e., affect each other in a causal way). In other words, as long as nothing can be transmitted between them faster than the speed of light, no laws have been violated.

Given that both the phrase "Big Bang" and the word "theory" have serious problems, you might wonder why astronomers coined the term to begin with. It turns out to have been a bit of a jest. In brief, in the mid 1900s there was a famous astronomer named Fred Hoyle who really disliked the idea that the universe might be expanding, because that expansion suggested the universe had a beginning, and if it had a beginning that suggested it might have been divinely created, which to Hoyle was unacceptable pseudoscience. The logic and irony here is fascinating, and Hoyle's position speaks to the power of confirmation bias. Discounting scientific evidence because it doesn't support what you believe is wholly antithetical to science. In a BBC radio interview on the subject, Hoyle pejoratively[6] referred to this idea of an expanding universe as a "Big Bang," which stuck.[7] So, in a nutshell, we are condemned to using the term "Big Bang" because a famous astronomer got on the radio and made fun of it, which goes a long way toward explaining why the phrase doesn't mean what many people think it does.

Another essential point that needs to be spelled out—scientifically, the "Big Bang" does *not* refer to actual creation of the universe to begin with, although the term is often conflated with this original instant of being. Rather the "Big Bang" solely refers to this period of radical inflation from whatever preexisting tiny nugget of a universe there might have been. Where that original nugget came from is another turtle down.

This is where the rubber hits the road as far as science goes: An idea as crazy as the universe expanding by a factor of 10^{26} in less than 10^{-32} seconds necessitates some compelling justification. To be sure, scientists didn't just wake up one morning and apropos of nothing say, "Hey, I have this radical idea the universe underwent extreme inflation 14 billion years ago." Rather, the idea of the Big Bang came about over many decades of slow burn, punctuated by a few dramatic revelations.

I do not want to go into a full and detailed history of the science that unfolded in the twentieth century that led up to the broad adoption of the Big Bang theory, which you can surely find a good account of elsewhere. However, because this book is (at least in part), about the limits of empirical inquiry, I do feel impelled to offer you a glimpse into some of these bits of empirical inquiry that convinced virtually all astronomers and physicists (and astrophysicists) to buy into this inflation hypothesis,[8] which even a hardened scientist must admit sounds kind of insane at first blush.

It turns out, we didn't even need to wait for modern astronomical observatories; I don't know if the fact that the night sky is dark has ever bothered you, but this turns out to be a nontrivial statement. At least as far back as the sixth century, the Greek monk Cosmas Indicopleustes was wrestling with the implication of having a dark night sky—intuiting that if the universe were infinite in time and space (and uniformly populated by stars), we would burn up.[9]

This observation is more generally known as the "dark sky paradox" but is often attributed to Heinrich Wilhelm Olbers as Olbers' paradox, an honor that science historians dispute his earning of.[10]

The dark sky paradox (which is, as usual, ultimately not an actual "paradox") goes like this: If the universe is infinite back in time, infinite in space, and approximately uniformly populated by an infinite number of stars, every line of sight out into space ought to land on a star. If this were true, the entire night sky should be bright and virtually uniformly blanketed by stars. Clearly the night sky is

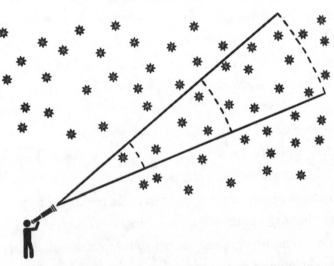

As we look farther away in the universe, individual stars appear fainter due to their distance, but there are more stars in our cone of vision as well. These two effects cancel each other out.

dark, which is obvious to even an untrained observer (unless, perhaps, they live in Manhattan), so one or more of these assumptions about the universe has to give.

My students often have their intuition tricked by their innate understanding that stars that are farther away will appear fainter to us, and surely this increasing faintness matters. It does, indeed, matter. But the catch is that each shell of the universe in our cone of vision should contain exactly proportionally more stars. In other words, the apparent brightness of the stars gets fainter as the distance squared, but the number of stars in the cone of our vision should also go up by the distance squared, and the two geometric effects cancel.

There are a range of solutions to the dark sky paradox, some of which are extremely contrived. But this "paradox" completely goes away if the universe had a beginning.

An important takeaway from the dark sky paradox is that we don't always need fancy equipment to come up with compelling hypotheses. Sometimes progress just requires asking a simple question about something we take for granted—like "Why is the night sky dark?"

Another "paradox" that burbled to the surface of physics in the 1900s was the fact that the universe had not decayed into statistical equilibrium. It was clear that applying the second law of thermodynamics to the universe resulted in a flagrant contradiction. Physicists at the time looked to the recently emerged theory of general relativity to argue that the universe was not subject to steady external conditions (i.e., it was not a closed system, as required for the second law of thermodynamics to be enforced), which provided a loophole to wiggle out of the apparent paradox.[11] To my mind, this justification just kicked the can down the road, leading to the questions of why the universe might not be subject to steady external conditions and what is external to the universe anyway. Be that as it may, the clearly obvious—even to a layperson—fact that the universe had not "run down" could not be ignored. This fact was part of the slowly simmering case that might suggest the universe was not, in fact, in a steady state.

Simultaneously, two other threads of math and astronomical observation were unfolding, which resulted in the discovery that the universe itself is expanding.[12] That is kind of a big deal.

I often try to imagine what it would have been like to have spent your life believing that the universe had been more or less eternally in its current state, and then be challenged to make this radical change to your worldview. Not surprisingly, there was major debate on this topic, which is a healthy thing when a potentially paradigm-shifting idea is introduced. But the data were irrefutable, and it became increasingly difficult to ignore that an expanding universe required that the universe was necessarily smaller in the past. When one followed this line of reasoning to its natural outcome, the inference was that the universe must have once been (perhaps) infinitely small. To be sure, there are other nuances that we should not be too hasty about sweeping under the rug—such as the tautology that we can only observe the region of the universe that we can observe; it could well be that there are distant regions of the universe doing radically different things. The good news is that the Big Bang hypothesis came in a nice package with very strong predictions that could be (and have been) observationally tested.

One strong and specific prediction from the Big Bang model is the precise relative abundances of elements in the universe. For us humans living on this rock with lots of heavy elements, you can be forgiven for thinking that things like oxygen and carbon are abundant in the universe. But once again, we have such a very limited view of reality. If you've never thought about where the various elements on the periodic table that haunted you in chemistry class came from, now is your chance. This is one of those *things that exist* that are easy to take for granted, like the darkness of the night sky. But when you dig a little deeper and start asking questions, suddenly new problems emerge. In this case, with all these elements on the periodic table, it seems terribly unfair that hydrogen gets roughly three-quarters of all the atomic mass in the universe today (and helium gets the remaining one-quarter). The other elements really got stiffed.

Back in the day, by which I mean within like the first 10^{-6} seconds after the Big Bang, we didn't even have any elements—rather just an exceptionally hot soup of elementary particles like quarks and gluons. Shortly thereafter (or "immediately" by modern time keeping), this soup cooled enough that protons, electrons, and neutrons could emerge. At that point, the universe only had about twenty minutes to create whatever elements it could before it cooled off too much to slam nuclei together with enough speed. Instead of taking a twenty-minute power nap, the universe squeezed in some element production, but all it had time to do in these twenty minutes was make some helium nuclei (by slamming hydrogen nuclei—in other words, protons—together) and an itty-bitty trace amount of lithium. At that point, the universe ran out of steam (but not literally, because we didn't have oxygen yet, so we didn't have H_2O), and those were the elements we had—for a very long time—until stars formed, which took a couple hundred million years. *All* the other elements were forged in the lives and deaths of stars. As the saying goes, you are literally made of stardust (although I would argue it is more like star guts, but that doesn't sound nearly as poetic).

I know that the physics of all of this might sound complex, but it is actually straightforward thermodynamics and particle physics. The point of which is that we can make very precise predictions about the exact abundances of hydrogen, helium, lithium, and even more importantly—their isotopes.[13] Chief among these isotopes is deuterium, or "heavy hydrogen" (because the nucleus has both a proton and a neutron). Unlike the other elements that are created in stars, deuterium is destroyed in stars, so as the universe gets older and stars burn through their fuel, the amount of deuterium is getting ever lower. Moreover, there are no known viable paths to create more than trace amounts of deuterium *other than* Big Bang nucleosynthesis. The ratio of deuterium to standard hydrogen (also known as "protium") provides exquisite constraints on the conditions and timescales of the first twenty minutes of the universe. And guess what, the data exactly support the Big Bang theory.

Another bit of evidence is admittedly circumstantial, but a challenge to explain without the Big Bang. If we look out toward the left 10 billion light-years toward the horizon,[14] and then we look to the right 10 billion light-years toward the other side of the horizon, the two regions (which were something like 20 billion light-years away from each other when their light headed our way)[15] appear to be statistically indistinguishable. In fact, every direction we look out to the horizon, the universe appears to be homogeneous.

If the universe had been sitting around in some kind of steady state, these regions cannot have been in causal contact and sharing information for 20 billion years, in which case the horizon has no business being homogeneous. Once again, cosmic inflation comes to the rescue. With a cosmic inflation scenario, the entire observable universe would have been in causal contact—at the end of inflation, today's observable universe would have been about a meter in diameter, or about 3×10^{-9} light seconds. In other words, there was loads of time for causal contact between different places in the universe when it had the size of a large dog. Just like that, the horizon problem goes away.

The Big Bang theory made one more key prediction that would be hard to reconcile with any other physical model: roughly 400,000 years after the Big Bang, the universe should have cooled enough to allow electrons to combine with protons and make bona fide atoms. For reasons I am totally glossing over (let's just call these reasons "physics"), when the protons and electrons are able to combine, light is able to stream freely out into the universe, and travel in whatever direction it was heading until it runs into something. This background source of light, which should be coming from every direction in the sky, has been dubbed the cosmic microwave background (or CMB). This CMB light should have a very specific observational signature that tells us exactly the temperature of the universe when it was emitted. Couple that with our knowledge of exactly what that temperature should be when the protons and electrons combined into atoms, and we have a very specific and testable prediction. The physics here is exceedingly well understood and easily done by hapless undergraduate physics majors as homework problems. What was less easy was testing this precise prediction.

Lucky for us, some of this light that escaped when electrons and protons combined into atoms proceeded to travel across the universe for more than 13 billion years and run into Earth. There have been several major experiments to measure the properties of the cosmic background, with steadily increasing precision. I think it will not surprise you at this point to learn that the observations exactly agree with predictions from the Big Bang theory. Again.

The bottom line is that there is now a mountain of evidence, using radically different predictions and experiments, that supports the Big Bang theory. The only other hypothesis to date that is 100 percent consistent with the experimental data is that a higher power contrived this universe as a set-up job to fake out us mere mortals into thinking there was a Big Bang. This hypothesis, however, notably lacks testability or predictive power.

Epochs of Confidence

While we have an extraordinary amount of evidence supporting *that* the Big Bang happened, that is not at all the same thing as understanding *how* or *why* it happened. As we consider eras in the cosmos before the present, as a general rule, our level of understanding diminishes. That being said, it may surprise you to hear that in many respects, our understanding of the physics in the universe—between the time when it was about twenty minutes old and 400,000 years old—is more complete than our understanding of the physical processes that have unfolded over the last billion or so years.

Our levels of confidence in what happened fall precipitously along with our ability to empirically test our theories. For example, the earliest time from which we can *ever* receive light is when the universe was ~400,000 years old—specifically the time when the universe cooled down enough for electrons and protons to settle down and make atoms. To be sure, there was plenty of light before that time, but it was trapped in a kind of particle soup. As a result, this is the earliest time period we can directly observe with light. Ever.

Just because we can't actually observe the universe at times before about 400,000 years doesn't mean we are totally in the dark. For example, our ability to detect gravitational waves is rapidly advancing, which will give us a whole new window on the universe. At the moment, our main tools to explore earlier times are fantastical experiments in physics labs. With modern equipment, it is commonplace to observe energetic conditions that would have been present when the universe was only a few minutes old, so the physics that unfolds under these conditions is well understood.

We are now astoundingly able to probe energy levels present when the universe was less than 10^{-32} seconds old, which is—at least to me even with my desensitized astrophysics brain—utterly mind-blowing. These insane energy levels can be achieved with the Large Hadron Collider in Europe. Please take a moment and let that sink in—modern experiments can explore the conditions of a 10^{-32}-second-old universe.

That being said, even with the Large Hadron Collider, we don't have unfettered access to these early times—the experiments are brief and limited. By analogy, our progress with the Large Hadron Collider is like trying to map the room around you by looking through a drinking straw and having no control over where the straw is pointed. Actually, never mind, it is a lot harder than that; the straw also needs to be out of operation frequently for repairs, and the room is dark most of the time.

At the moment, we have no window—observationally or experimentally—into the universe at times earlier than the Large Hadron Collider can recreate the energetics of. This includes the period of cosmic inflation. We have a working hypothesis for what might have caused inflation, which involves states called "false vacuums" and something called an "inflation field," which colluded to generate negative pressure, which in turn triggered the incredible cosmic inflation. But, as of the time I'm writing this, we haven't been able to go back 13.7 billion years to test this hypothesis, and we also haven't (yet) recreated these conditions in the lab.

The cause of the inflation is still an unsolved mystery in its own right. But—and this is an important "but"—I think this is one of those mysteries that might actually be solvable soonish—maybe even with the next generation of supercolliders.

When we try to think about even earlier times, like, say, the Planck time of ~10^{-43} seconds, we are now in a real pickle because even the laws of physics as we currently understand them break down; a consequence of Heisenberg's uncertainty principle is that any measure of time smaller than the Planck time is indeterminant. Similar to the situation we will encounter at the heart of black holes, we confront an apparent singularity at the creation of the universe, and the only way out seems to be a theory of quantum gravity.

That being said, thinking about what generated cosmic inflation—or even what was going on at the Planck time—isn't what keeps many people up at night thinking about what caused the Big Bang. Rather, the deeper conundrum arises when we go a turtle down and consider why there was even *something*—however compact or bizarre—to be inflated to begin with. Or in other words, why is there something instead of nothing?

Something from Nothing

I don't know how you feel about the idea that the universe came from "nothing," but this plagues me. This is the root of our bewilderment—whether there can be a "nothing" from which "something" is generated. This question truly keeps me awake many nights, but I draw some solace from knowing that thinkers have wrestled with this topic for perhaps as long as there have been thinkers to wrestle with it. *Creatio ex nihilo* (creation out of nothing) has a venerable history; Thomas Aquinas, for example, considered this question to inherently offer proof of a god as a "first cause."[16]

If you have felt stymied by this "something from nothing" question, at least you can know that you are not alone.

When I was a graduate student, I had the impression that the solution to this philosophical knot was obvious to others, and I just didn't get the right memo. The party line in physics seemed to be a version of "The birth of the universe came about from a quantum fluctuation," and that was that. To be sure, this explanation is still very much on the table, and we will come back to it in more detail. However, what always bothered me about the cursory response to how the universe came into being was an implicit avoidance of what was (at least to me) the deeper question—fine, let's say a quantum fluctuation caused the universe to come into being, but what enabled a framework for quantum mechanics to even exist and for the laws of physics to have an arena in which to operate? Or put another way, what is the realm in which these laws that govern the behavior of everything live? I kept these questions to myself for a very long time, but they didn't go away.

Many physicists will argue (and have argued with me) that there is no point in trying to go another turtle down past the cause of inflation. This apparently philosophical question of why there is something instead of nothing frequently exasperates many folks as "outside the realm of science." I will concede that they may have a point, and it may truly be outside the realm of science. If this question is ultimately unanswerable, perhaps dwelling on it has no practical value beyond making astrophysicists like me even more sleep deprived. However, I would also argue this avoidance position becomes self-fulfilling; if scientists just give up and stop studying something because there might not be a scientific explanation, then we have very little chance indeed of coming up with a scientific explanation. Somehow knowing the possible futility is not sufficient to deter my mind from going down this rabbit hole, and I don't think I'm alone.

I also believe that not engaging in the "something-from-nothing" issue is a bit of a cop-out—science itself has led us to the precipice of perhaps the greatest existential mystery that has impelled human thought and belief for uncounted generations. To be sure, there is currently a handover that must happen between science, philosophy, and theology somewhere around 10^{-43} seconds, but I don't think scientists are totally off the hook, because there may well be another turtle that is accessible to empirical inquiry of the future.

Before we can get into if or how one might generate "something" from "nothing," we need to take a step back and talk about "nothing" with more precision. When I teach a course on this, we spend an entire class period talking about "nothing," for which I am sure there are cynics out there questioning the use of tuition dollars. However, the thing is, there are different types of "nothing." In fact, many types of nothing have been expounded on in academic discussions, but I will limit us to three main types here:

The most trivial type of nothing in this scheme is Type 1, which is what I suspect many people first think of when they think of the concept of "nothing." In this case, we have a turnkey universe, there just isn't anything in it—no matter, no energy, no force fields. To be clear, we can't even achieve this type of nothing in a lab—even in the most extreme ultravacuum chambers, there are still fields that permeate space that we can't avoid (at least not yet), such as the Higgs field, which is responsible for giving mass to things.

Type of nothing	Matter/Energy exists	Space-time exists	Laws of physics exist
1	∅	✓	✓
2	∅	∅	✓
3	∅	∅	∅

Type 2 nothing turns nothingness up a notch—in this case we no longer have the very fabric of space-time. In a real estate analogy, it would be like having a permit to build a house, but not yet even having the land to build it on. There are, however, a whole mess of zoning regulations that determine what can and cannot be done. This type of nothing no longer has "things" in it or an arena in which there even could be things, but underlying laws "exist" that inform what is possible. This type of nothing overtly raises the question of *where* these laws reside if they don't have a universe to reside in, but this issue extends beyond this type of nothing (and which we will come back to in the chapter on the laws of the universe).

One can imagine a higher power in the context of Type 2 nothing, and they might determine what is and is not possible—in other words the laws that govern the universe reside with them. This is indeed the perspective of many theologies and may well be correct for all I know. But we have only managed to nudge the philosophical can down the road—even a higher power is "something," so we still have something instead of nothing.

The most extreme type of nothing is Type 3. No things, no fields, no arena, and no underlying framework or set of laws. Truly a complete and utter void that defies language—even calling it a "void" might suggest that there is *something* that is empty. This type of nothing is the most challenging to reconcile with there now (demonstrably) being something for us to make for dinner. In this type of nothing, it is not clear there is even room for a higher power—because that would not be nothing.

If you are expecting me to resolve this philosophical quandary, I'm afraid your faith is sorely misplaced. I'm just as confused as anyone. I have (mostly) come to terms with only being able to go one more turtle down at a time, and then hold on to some hope that perhaps when we have understood that next turtle things will be more clear.

Before we go on to speculate about what might be one turtle down, I want to pause and bring up a curious feature of the observable universe. Let's imagine for a moment that you want to buy a car, but you have no money. A very kind and generous friend loans you the money with no interest, and you buy the car. Now you have the car, but you owe your friend exactly the amount of money the car is worth. So, in some very real sense your net worth is zero. It turns out the universe is in a similar situation, only the loan was a lot more extreme.

The "net worth" of the observable universe appears to be effectively zero (or as close to zero as we can determine). We can estimate the entire mass/energy of the observable universe (which is positive), and we can estimate the entire gravitational potential energy of the universe (which is negative), and these two seem to cancel each other out. If the universe has a net worth of nothing, then instead of asking, "Why do we have something instead of nothing?," the question becomes "What was the mechanism for the loan?"

We are clearly on the borderlands of science here, but that hasn't kept humans from thinking about what might have preceded[17] the universe as we know it, even if these conjectures are (currently) outside the realm of empirical inquiry. The general tactic is to consider the physics of the universe as we understand it and try to extrapolate possibilities.

One concept with a long history is that the universe is created and destroyed in a cycle. The idea of a cyclic universe has been present in various cultures and mythologies throughout history. For example, in Hindu and Buddhist cosmogonies, there are the concepts of *kalpa* and *saṃsāra*, which refer to cycles of creation, destruction, and rebirth of the universe.[18] Ancient Greek cosmogonies are rife with notions of cycles; Empedocles (fifth century BC) developed a cyclic cosmogony that even included the concepts of love and strife.[19]

A cyclic universe understandably has broad appeal—it nicely skirts the issue of having a "beginning," and there is an important psychological element to believing there will be another cycle that will start anew, no matter how bad things might seem now. The general modern idea of a cyclic universe proposes that the universe undergoes an infinite series of cycles, with each cycle beginning with a "big bang" and ending with a "big crunch." According to this hypothesis, the universe expands and cools down after the big bang, eventually slowing down and collapsing in on itself due to gravity, only to rebound and expand again.

This was all well and good and seemed totally plausible until we went out to empirically test how fast the expansion of the universe was slowing down. This led to the discovery of dark energy (I am confident in your ability to turn pages and go find the chapter on dark energy, so I will blatantly skip over details here). To be certain, there is *a lot* we do not understand about dark energy, but insofar as we understand anything about it, we know it is getting stronger with time and the expansion of the universe is accelerating. Unless there is some type of phase transition in the cards (which I absolutely do not rule out), we have no cause to think this acceleration will stop. In which case, a cyclic universe is pretty much out of the question, which is kind of a bummer.

Among physicists, the most popular explanation for the origin of the universe may be quantum mechanics. One of the great things about quantum mechanics is that it gives us a lot of loopholes to play with in terms of what is "allowed" to happen, which provides nearly unlimited fodder for pseudoscientific conjectures. Nevertheless, quantum mechanics itself is well-established and empirically verified. One aspect of the quantum world is that concepts like "cause" and "effect" are not even rigidly defined. This decoupling of cause and effect is like a breath of fresh air in a room full of hypotheses about what "caused" the universe to be created (especially if time didn't exist)—giving us this cool trick to conjecture that perhaps there wasn't a "cause" at all, because that very word loses its meaning.

Another key feature of quantum mechanics (one that we will encounter in more depth in the chapter on black holes) is that particles can and do pop in and out of existence from "nothing," which is required by the very nature of the uncertainty principle—in this case, energy and time can't simultaneously have precise values, and "nothing" has a very precise value of 0. I will grant you that there is a world of difference between a *particle* popping into existence and an entire *universe* popping into existence, but the loophole is there. This does bring up another issue, though—for anything to pop into existence from a quantum field, there must be an underlying quantum field to begin with. So, what kind of "nothing" is that?

Of the various possible explanations for the "beginning" of the universe, this next candidate is perhaps the most abstract. The "no-boundary" hypothesis requires fundamentally thinking about time and space in a way that is far removed from your normal daily experience. This hypothesis was originally put forward by Stephen Hawking and James Hartle in the 1980s with an explicit goal of avoiding a "beginning" of time, as a result, in this proposal time has no beginning.[20]

The concept here is a bit abstract, so let's turn to another analogy.

If you will, imagine a standard globe of the Earth with all the usual lines of latitude and longitude. Now further imagine you happen to be an intrepid adventurer (or perhaps you actually are and you don't have to imagine), and you are headed to the South Pole. As you travel south, you are of course aware—as any intrepid adventurer would be—that the lines of latitude and lines of longitude are perpendicular, which is handy if you are trying to discern between the axes of north-south and east-west. Or at least they are perpendicular until you get to the South Pole, at which point something odd happens—every direction you might face is north. Not only is there no "south" anymore, but "east" and "west" have the same problem. Of course, you—standing at the South Pole—don't perceive anything particularly different about the precise South Pole, as opposed to a few meters in either direction. This collapse of all directions to the "north" is purely a result of the standard coordinate system we have in place to describe locations on Earth.

In a nutshell, this is how the no-boundary universe proposal works: the basic idea is that time and space are orthogonal dimensions in space-time (think of lines of latitude and longitude on the equator), with time dimension effectively oriented along the north-south axis. But the trick of this proposal is that by the time we get to the "South Pole" of the universe (aka the beginning of time),

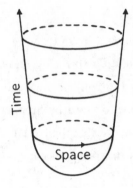

A schematic of the no-boundary universe conceived of by Stephen Hawking and James Hartle. In this geometry, the location where time is 0 is akin to the South Pole of the global coordinate system on Earth; there is nothing farther south than the South Pole. Similarly, in this scheme, there is no time before time = 0.

there is no further south to go. If time = 0 is at the South Pole, then there is no such thing as time = −1, and therefore there is no "before" in this coordinate system.

Another way to think about the nature of time in this scheme is that time becomes "imaginary" or alternately that time becomes more space-like. To envision this, if you thinly sliced up the globe along lines of latitude you would end up with a stack of circles growing and shrinking in size from north to south. These circle slices are slices of space. At the South Pole, not only does the size of this slice of space become vanishingly small, but the dimension of time, which started out as orthogonal to these slices, has curved into the same plane as the dimension of space.

This proposal of a no-boundary universe neatly gets around the standard notion of causality and what came "before" the universe because it avoids the concept of "before" altogether. However, it notably doesn't explain why we have an arena in which this universe can unfold to begin with.

Coming out of the gate, this next scenario might sound like pure science fiction, but it is tethered to real physics that is being actively pursued (and on which legitimate peer-reviewed papers are penned).[21]

This scenario is an alternative to inflation, but I include it here because it also speaks to the "What happened before the Big Bang?" question.

The gist of this scenario is that our universe could be a multi-dimensional "brane" (short for "membrane," which is traditionally only two dimensional) that lives in a volume of yet one extra dimension (and because we are so good at naming things, we call this the "bulk"). By analogy, I like to think of big flat pieces of sea kelp floating in the ocean—where our universe is the kelp and the bulk is akin to the water—which is a peaceful counterpoint to what is happening here. The key is that our brane is not the only one, in fact there may be an infinite number for all we know (which is very little). If two of these universe-branes collide, a whole ridiculous amount of energy is injected into those brane-universes. One of the features of this scenario is that it doesn't have a time = 0, thus there is no "beginning" of time, rather there is a possibly infinite time in this metaverse in which branes collide and move apart.

I warned you that it sounded like science fiction. The thing is, though, that there is real physics that provides just enough circumstantial evidence for us to not abandon this theory. The whole crazy idea of colliding branes is fundamentally grounded in another possibly-crazy-but-beautiful idea called "string theory," which we will talk more about in the chapter on other dimensions. Isn't it annoying how the topics of these chapters get intertwined? Believe me, I hear you. Imagine trying to figure out in what order to present things in this book to begin with. But the universe is complicated and entangled that way, and perhaps we shouldn't be surprised that many of the unsolved mysteries are connected.

The good news is that this so-called "ekpyrotic" scenario makes some empirically testable predictions that differ from the standard inflationary paradigm, including predictions for the spectrum of primordial gravitational waves that have not yet been detected. My read of the current research status is that the results, so far, lean toward inflation, but I am not quite ready to close the book on this crazy ekpyrotic scenario. Heck, even if colliding branes were not responsible for the Big Bang, not-colliding branes could still be a valid formulation of reality.

But where did the branes come from to begin with?

The Big Bang and black holes are the two astrophysical environments in which we encounter annoying singularities and the laws of physics as we understand them get into hot water. This shared property brings us to the next possible origin for the universe.

You may also note with intense curiosity that these two environments kinda sorta seem like opposites—for example, space itself flows *into* black holes, but *out* of the Big Bang. In fact, doesn't the Big Bang itself seem exactly what one might expect from the opposite of a black hole—a *white* hole? Yep. In fact, this has an actual name—black hole cosmology, and there are once again legitimate peer-reviewed papers written about it.

There is another nifty fact to add relating to dark energy. In our current (admittedly limited) understanding of dark energy, it acts like there is a *negative* pressure on the fabric of space, and in Einstein's field equations, negative pressure acts like a repulsive gravity. It just so happens that one way to get negative pressure is by stretching the fabric of space—and hey, guess what happens inside black holes? It is entirely within the realm of possibility (and is likely to remain so for a good while since we are a long way from knowing what happens inside black holes) that deep inside the black hole space itself is stretched to such an extreme that dark energy takes over—inflating space like a balloon. Sound familiar?

As far as we can tell, our observations are consistent with us living in a universe created inside of a supermassive black hole formed in a parent universe. That is the good news. The bad news is that we don't currently have a way to empirically test this and discriminate between living in a black hole and living in a conventional (and also not currently well understood) inflationary universe.

This hypothesis is tantalizing to think about, though, and if I got to pick which one of the solutions I *want* to explain the Big Bang, this would be it (sadly, reality doesn't really care what I want, or what you want for that matter). One of the side effects of black-hole-baby-universe conjecture is that our own universe could have untold offspring. Moreover, it might be possible for parent universes to pass along a form of genetics through the precise values of physical constants that get passed down,[22] which could come from an aspect of string theory (in which precise properties of the universe are determined from the shape of compactified extra dimensions). If genetics are passed down, this in turn could lead to cosmological natural selection, as proposed by physicist Lee Smolin: the more well-tuned the properties in a universe are for making black holes, the more baby universes it will have.[23]

Need I point out that we still have to contend with where the first parent universe came from? That is at least another turtle down.

We can't talk about the origin of the universe and not bring up the possibility of a higher being, if for no other reason than a huge fraction of the global population believes this to be true. Just as we can't prove there is a god, we also can't prove there isn't, so I find that I have limited patience with explicit hard-core atheists (by which I mean people who reject the idea of a higher power as impossible, not people who simply lack a belief in the concept and lean more toward agnosticism). Given that we *know* there is so much we don't know about the cosmos (let alone the things we don't even know we don't know), and the reality that our senses and human intellect have serious limitations, I think it is an ultimate show of human hubris to outright reject the possibility of things we don't understand.

To be clear, in this context I am not necessarily talking about a conventional "god" as I think might be commonly envisioned. Really any sort of higher power that resides outside our sphere of reality might do. This could be some sort of superintelligent being—at least superintelligent by our standards, though they could be profoundly idiotic among their own kind, but I hope for our sake this is not the case. All of the cosmos could even be an experiment, or part of a higher-power school science project. When I was in elementary school, I did an experiment with a set of plants that I suspect is fairly common—some plants got water, some didn't. Some got sunlight, some didn't. Some had bleach poured on them, some didn't. Who knows, maybe our cosmos is someone's homework.

And yet again, a "higher power" is not nothing.

Magical Thinking

As far as mysteries of the universe go, thinking about the origin of the universe sends me into more of an existential crisis than any other topic. Not only are the physics of this uncharted (and potentially unchartable?), but I end up tangled in philosophical knots. It is clear to me that my own intellect is not up to the task—there is an inherent problem embedded in the infinite regression that has no apparent solution that I can convince myself of. For this reason, I am often tempted to punt this issue as beyond human comprehension, and therefore into the realm of the supernatural.

Then another little voice in my mind speaks up to remind me that just because I don't understand something, doesn't mean it isn't understandable. I am reasonably sure that no matter how hard I might try, there is just no way I could teach my dogs to understand general relativity. Heck, even thermodynamics would be a stretch—once I brought helium-filled balloons home for a party, and one of my cats at the time absolutely lost his mind in terror. It seemed obvious to my husband and me that Jinko's feline brain was thinking along the lines of "Those things that float around the house violate the laws of nature."

Here we are with our limited human brains—do we really have so much arrogance as to think just because we don't understand something it must be supernatural? This is a good time to remind ourselves of the "God of the gaps" fallacy from the last chapter on epistemology. Of course, the inverse is also true—just because we think we understand something, doesn't require that it not be supernatural. Also recall the third of Clarke's three "laws": "Any sufficiently advanced technology is indistinguishable from magic."

Interstition

One question that often follows closely after "Why is there something instead of nothing?" is why this *something* happens to include life. This question predates modern science but may now be within our reach.

A confirmation of extraterrestrial (ET) life would arguably be the most profound discovery of human history—bigger even than fire, the wheel, or even sliced bread. Such a discovery would not necessarily impact our fundamental understandings in modern science, but it would undoubtedly affect how we see ourselves and our place in the universe.

The religion I grew up in forbade the existence of extraterrestrial life, which even at the time I thought was an odd thing to oppose. I was pointed to the Bible and given the argument that the universe was made for humans. I suppose if one doesn't understand how expansive the universe is and thinks they are the center of it, the notion that humans are the point of it all doesn't seem as ludicrous as it is.

4

Does Extraterrestrial Life Exist?

EXTRATERRESTRIAL LIFE. WHAT IMAGE POPS INTO YOUR mind when you read that phrase?

If you're like most people I've asked this question of, you probably envision something along the lines of a character from a movie or TV show that is, for all practical purposes, essentially humanoid. Perhaps your version of ET life has green skin or eight arms or a tail, but these characteristics are not terribly different from the basic physiology of people. Thanks mostly to Hollywood, people often have a shockingly uncreative image of what actual ET life might look like. If you did not envision a Hollywood-ish type character, give yourself a pat on the back with one of your likely two (and not eight) arms.

Whatever you might have thought of in the previous paragraph, why do you think *that particular example* is what you thought of and not something else? Even in our own biosphere on Earth, there are life-forms that are far more "alien" looking than many people think of at the mention of "ET life." The ocean, for example, has forms of life that are much stranger than in most movies or TV shows.

A plush dumbo octopus would be cute, and if we ran into one in space, we might want to hug it (for the record, I have no idea how bad of an idea that would be). But coming across a human-sized amphipod in a dark space alley would be the thing of nightmares.

One of my favorite life-forms from Earth was once thought to be our great-great-great-great-(repeat for a long time) cousin, the *Saccorhytus coronarius*. This critter doesn't exist now but appears to be a 540-million-year-ago example of life in the Cambrian era and looked more "alien" than almost anything in Hollywood. You might note that, among other interesting characteristics, *Saccorhytus coronarius* is not believed to have had an anus. There are so many jokes waiting to be made, and I am glad for its sake that middle school kids weren't around at the same time.

Images of (*left*) *Saccorhytus coronarius* (based on visualization of Nobu Tamura, CC BY-SA 4.0); (*middle*) a dumbo octopus; and (*right*) an amphipod.

The point is that even on Earth we have a variety of life that calls into question whether ET life—in totally different environments—would be remotely humanoid. In fact, most of the life-forms that might come to mind are fairly macroscopic—but these "big" forms of life are *by far* in the minority. And speaking of microscopic forms of life—it's worth considering to what extent your body is really *your* body. Fun fact: there are more bacterial cells in your body than human cells (roughly 38 trillion compared to 30 trillion). So, who are you really?

Consider again the ET life-form you thought of at the beginning of this section: What aspects of its environment would have caused it to evolve that way? If you end up basically recreating the surface of Earth, I challenge you to push your thinking further; think of a planet very different from Earth. What characteristics might life need to evolve and survive in this radically different environment? Or what about life not on a planet at all? There are good reasons to start by using what we know about life on Earth to envision life elsewhere. But there are also good reasons to step back and consider that life might develop in the universe under radically different conditions. If we're asking whether there is any ET life, we need to know what we mean by "life," and we must be open to the idea that "life" somewhere else in the universe might be extremely different than life on Earth.

Being "alive" is one of those things that it is tempting to say, "I know it when I see it." But are you sure? Trying to actually define life is a bit tricky. The renown physicist Erwin Schrödinger attempted to define "life" in terms of physics in his 1944 book *What Is Life? The Physical Aspect of the Living Cell*. Schrödinger argued that life is defined by its ability to resist decay to equilibrium (or in more physics-like terms, to resist "entropy"). We are going to talk a whole lot more about entropy later. Try not to let this physics-y term freak you out, it really isn't that bad, and it has huge implications for the universe. For now, I will just leave the term here and not derail your reading about ET life with a tangent about physics.

This is an interesting working definition, but there are subtleties—for example, over what time period should we say something has to resist decay? If we want to consider ourselves to be alive, and most people I know are fond of this idea, we must acknowledge that as we age, we decay, and once we are dead, we decay a whole lot more until—eventually—we reach equilibrium with our environment again. The point is that we don't 100 percent resist decay. In fact, you are decaying at this very moment.

Even biologists debate the definition of life, so if you are having trouble coming up with a robust definition, you shouldn't feel too bad. A general working set of requirements for something to be considered "alive" might include some subset of the list below. But it only takes a few questions to throw a wrench in the mix.

Is composed of cells. What exactly do we mean by "cell"? Does it have to be the straight-up biologically understood definition? Why? That seems awfully narrow-minded of us.

Has a metabolism (uses energy). One might argue that rocks don't use energy, but what about refrigerators? Or computers? Does it have to use energy for a purpose (if so, who gets to decide what counts as a "purpose")?

Can grow. I don't *a priori* see why growing is required for being alive. And what about a person who is no longer capable of growing? We're in big trouble if people are only considered "alive" through their adolescent growth period (which I say with some degree of bias as a middle-aged woman).

Can adapt. What range of conditions does something need to adapt to? Humans can adapt to a small range of conditions, but we have our limits. Some people I know are really pretty terrible at adapting to pretty much anything.

Can respond to stimuli. Responding to stimuli is interesting because of physics. There is this whole Newton's third law thing—actions and reactions, which means, at some level, that everything in the universe meets this criterion. For the sake of argument, let's assume "responding" is more nuanced in some way. (What way?) Does an active choice have to be involved? In which case, what about nonconscious forms of life, like plants? And what if we don't have free will to begin with?

Can reproduce and display heredity. Lots of things that normal people would generally consider to be alive are not able to reproduce. Take mules, for example, which are the result of a donkey and a horse being special friends. Donkeys have sixty-two chromosomes, horses have sixty-four, and mules end up with sixty-three—resulting in mules being infertile (OK, to be fair, there have been a few confirmed cases of a mule giving birth—but it is exceedingly rare). Just because someone can't have offspring, would that mean they are not alive? Or do we only need the criterion to be statistically met by the particular class of things, while individuals can fail?

Maintains homeostasis (stable internal conditions). Homeostasis is a big one. If you can't maintain homeostasis, you die. Well, at least according to our working definition of life (which is admittedly pretty crummy). Humans can maintain homeostasis through a limited range of environmental conditions (pressure, temperature, chemical composition, etc.), but we are not actually all that robust as far as life-forms go. Drop someone in an Arctic winter without shelter, and they are not likely to be able to maintain a "stable internal condition" for very long. What about computers? They have fans that turn on if they start to overheat. Does that count? What about the Earth itself, which has numerous processes that *try* to keep the planet in homeostasis, like the geological carbon cycle, which takes millions of years (and which we humans have screwed up in just a couple centuries).

Can evolve. What exactly do we mean by "evolve"? Is it just that each generation is different than the last? (Which requires reproduction, by the way.) Or does each generation need to be "better" in some way? (Who gets to decide what is "better"?) What if a life-form reached a hypothetical apex of evolution and there were no more "evolved" it could get? The very presence of this criterion also means that we can't define something as "alive" unless we can see what its kin have done over multiple generations.

Can die. The ability to die puts us straight into a semantic loop, because to most people dead = not alive (Schrödinger's cat aside). To say whether something is "dead," we need to be able to define what it means to be "alive." What if (*hypothetically*) in the science of the future we found a path to immortality. Would that mean that our future immortal offspring would not be alive?

I hope you are feeling deeply unsatisfied with any working definition of life. If we break this list down, we find we need to keep making exceptions under different circumstances. Do *all* these criteria need to be met? If not, how many of them are necessary? Does it matter over what range of times or conditions the criteria are met? If there are enough "exceptions" to the rule, then eventually the "rule" has very little value. Settling for "we will know life when we see it" is not particularly satisfactory. My sense is that we may eventually need to establish a continuum of "aliveness," from, say, zero to 10. I'm sure that we would consider ourselves to be the apex of living and rank ourselves a 10 on this scale, with typical human hubris.

Life-as-We-Know-It (e.g., Life on Earth)

We find ourselves in a quandary. We know that we don't know everything there is to know about possible life in the universe, but we must start somewhere. Even just in our own biosphere, we have some spectacular life-forms that can survive in crazy conditions. We have found life living (and flourishing) in conditions that are hostile to human existence, which has implications for physical conditions in which life-as-we-know-it is possible.

Some of the critters that can withstand (and even flourish) in hostile-to-human conditions firmly land in the category of creepy-crawlies, like special kinds of worms. I grew up learning that without sunlight we wouldn't have life. It turns out that is not quite true, and giant tube worms are a great example of why. In the case of these worms, they get their energy by living on and near undersea thermal vents. There are also methane ice worms. These inch-ish-long critters have an ideal vacation spot in the deep, cold, dark ocean, buried in toxic (to humans) material. Like the giant tube worms, these ice worms don't use light from the sun as part of their ecosystem; instead, they prefer to feast on bacteria that, in turn, feast on mounds of methane ice.

On the other end of the spectrum of temperature from methane ice worms, we have *Pyrococcus furiosus*—even the name alone sounds like it should be part of a grimoire and translates roughly into "raging fireball." That should give you a solid hint at their superpower. The ideal temperature for these little beasts is 100°C or 212°F. Think about that the next time you are hoping to sterilize water for drinking by boiling it.

Another physical condition that is bad for people is exposure to high-energy radiation, which is a big concern in human space flight. But once again, humans are relative wimps in this respect. Enter *Deinococcus radiodurans*, which literally means "dreadful berry that withstands radiation."[1] Fortunately, it has a nickname that you can use instead—Conan the Bacterium, because it is so freaking tough. Conan the B. is so badass that it actually holds a place in *Guinness World Records* as the toughest known bacterium. It can survive all kinds of nasty things like dehydration, extreme cold, acidic environments, and the vacuum of outer space (making it a polyextremophile). But where Conan the B. really leaves the competition in the dust is its radiation resistance. Conan the B. can withstand one thousand times more radiation than humans with essentially no damage.

Another trick that might be helpful for life in space is suspended animation (cryptobiosis), which is deployed by lots of critters, including bdelloid rotifers, which can go into a state of cryptobiosis for tens of thousands of years and come back to life like nothing happened and wonder what's for lunch. They are only one example of something that has survived below the permafrost in Siberia and come back to life when thawed. What could possibly go wrong with global warming?

Finally, you just can't talk about extremophiles and not talk about tardigrades, and I saved them for last because they're my favorite (I even have a pair of plush tardigrade slippers gifted to me by friends). As far as polyextremophiles go, these are at the top of my list. They can survive temps close to absolute zero and more than 150°C or 300°F. You can hit them with up to almost a thousand times more radiation than humans, and they'll walk away (literally, because they have cute little squishy legs). They can also withstand the vacuum of space and at least a decade without water. What's not to love? I've even found some living on lichen near my house and kept them as "pets" for a time.

The point is that it's clear that even life-as-we-know-it can survive—and even flourish—in a range of conditions. As far as life-as-we-know-it, there are *some* physical conditions that seem like requirements. We think. As far as we can tell, most (all?) life on Earth needs a few basic things. But then again, if these conditions didn't exist, we can't say with certainty that life wouldn't have evolved and adapted with different requirements.

Perhaps the top requirement for life-as-we-know-it is some type of energy source. As we saw with the methane ice worms, there are sources of energy other than the Sun that can and do also support life, including thermal energy and chemical reactions. If we really want to think outside the box, we could even try to envision some form of life-*not*-as-we-know-it tapping into radically diverse sources of energy like magnetic fields or gravitational waves. There are a vast range of energy sources in the universe that could be harnessed in creative ways.

For life-as-we-know-it, complex molecules also seem essential, and these complex molecules require heavy elements. When the universe first kicked off, hydrogen and helium were the only elements around—until stars started dying and churning out heavier elements forged in their cores. Hydrogen and helium are great, and we owe them a debt of gratitude, but they suck at making complex molecules. Carbon is vital—the fact that carbon can make such complex molecules is what makes organic chemistry so infamous for its difficulty. Silicon, which sits right under carbon in the periodic table, can also make complex molecules, but the bonds aren't as strong, and as a result, molecules based on silicon are more fragile. The need for complex molecules suggests that life-as-we-know-it could not have emerged in the universe until enough stars had enough time to live and die and distribute their heavy elements around. As a reminder, you are literally made of these elements that were churned out of stars over billions of years.

When talking about life in the universe, you'll also hear a lot about liquid water, and for good reasons. Liquid water is essential for facilitating chemical reactions and as a transport medium in biological systems. The water molecule has some astounding properties that I think are woefully underappreciated. For starters, because of the structure of the water molecule—with the oxygen and hydrogen atoms not being in a straight line, it is slightly negative on one side and slightly positive on the other. This property results in water being a "universal" solvent. To be fair, it isn't actually "universal" and can't dissolve everything (thank goodness, otherwise showers would be a very different experience), but it dissolves more things than virtually any other liquid. Dissolving things might sound bad (I'm picturing some gruesome interaction with battery acid), but water's ability to dissolve things makes it a workhorse in distributing important nutrients.

You may well take the fact that water expands when it freezes for granted, given how common this phenomenon is in our modern lives. But this property is rare. There are a few other materials that will expand when frozen, but generally stuff gets smaller and denser when in solid form. What does this have to do with life? If water got denser when it froze, lakes, rivers, oceans, and so on would freeze from the bottom up until they became frozen solid. However, because ice is less dense than water, it floats on the surface, not only acting as insulation to the underlying water, but fundamentally enabling life to continue below the nice crusty layer of ice. Without this quirky property of water, life on Earth would be very different indeed (and possibly nonexistent).

As far as liquid water goes, the distance of the Earth from the Sun is right smack in what is called the "habitable zone" or the "Goldilocks zone"—meaning we are just at the right distance to have liquid water. A little farther away and water would freeze. A little closer and the liquid water would evaporate. Curiously, Mars is dancing right on the edge of this habitable zone, and there is compelling geological evidence that Mars once had rivers, lakes, and perhaps an ancient ocean. Today there is enough known water ice on Mars to cover the entire surface by 35 meters, and there is likely more hidden away deep beneath the surface. It is entirely possible that life emerged on Mars at some point and is maybe even still there. This possibility has led some to consider attempting to "terraform" Mars to make it suitable for human habitation. However, for the record, terraforming Mars to be "suitable" for humans by nuking the surface (as a well-known billionaire entrepreneur is promoting) is a terrible idea for all sorts of reasons, including: (1) it won't work, and (2) it is ethically abhorrent.

However, being in a habitable zone is not the only way to have liquid water because (once again for the kids in the back) sunlight isn't the only source of energy around: if there is frozen water and a sufficient source of energy, liquid water is inevitable. Take Europa, for example, which is one the plethora of moons of Jupiter (you can be forgiven if you don't remember them all, neither do I) and way outside the nominal "habitable zone" in the solar system. Europa has a thick shell of ice, underneath which is a mineral-rich subsurface ocean. Kind of like an M&M of planetoids. Europa, for example, gets a whole lot of energy through this whacky phenomenon called "tidal heating." As Europa orbits Jupiter, its insides get squished around from Jupiter's gravity (not unlike the tides on Earth, except our tides are on the surface). All this squishing generates friction, and friction means heat. And *voilà*! Liquid water!

While having some type of liquid seems important for life, and water seems especially well-suited to the task, it is less clear whether that liquid *must* be water. For example, we know there are lakes of methane elsewhere in the solar system, and we also know there are some extremophiles (e.g., methane ice worms) that don't mind a lot of methane. So that begs the question of whether some type of life could survive, adapt, and even flourish with an alternate liquid. Or indeed—if we are really pushing the envelope—whether the necessary chemical reactions and transport could happen in some way without any liquid at all.

Finally, another requirement for life to get going is time. Life is going to take a beat to evolve from simple organic compounds, and it needs a nice, comfy, safe place to really start a family. Exactly how long life needs to get started is open for debate, but there is some tantalizing circumstantial evidence to consider. To lay out the case, we need to go way back to the early solar system shortly after it started to assemble. The solar system was a nasty place back then—things weren't going in nice, orderly, circular-ish orbits yet, and stuff was flying around smashing unsuspecting rocks to bits. The situation got extra intense about 4.1 to 3.8 billion years ago, when evidence suggests that the gas giant planets changed orbits (technically called "planetary migration"), which unleashed dynamic chaos on the whole solar system. This period was called the "Late Heavy Bombardment" because it was "late" in the early evolution of the solar system, and the bombardment was indeed quite heavy. If you can envision a rain of planet-altering asteroid impacts, you're on the right track.

It's possible that some form of life emerged on Earth before the Late Heavy Bombardment, but we will probably never know because the entire planet would have been catastrophically resurfaced, almost certainly annihilating any existing life. Perhaps several times. This is where the situation reaches a fresh new level of interesting as far as the "how much time is needed for life" discussion goes, because the earliest evidence for life on Earth via chemical signatures dates from roughly 3.6 to 3.8 billion years ago—*right after* the late heavy bombardment ended. If you want something more substantial than chemical signatures, the earliest fossil records are from about 3.5 billion years ago, which on astronomical timescales is just a heartbeat later.

These timescales indicate that life on Earth emerged basically as soon as it possibly could have. In turn, the apparent speed with which life emerged might *suggest* that it was relatively easy for life to get going. On the other hand, given that we only have a data sample of one planet with life emerging (so far) we need to be a tad cautious in drawing conclusions. But if we take the existing evidence at face value—because it is the only data we have—it tells us that sometimes (whether "sometimes" is closer to "commonly" or "rarely") life isn't so hard to kick off.

There is one more key obstacle to proto-life having enough time in a nice, comfy, safe environment, which is the lifetime of the central star in a given stellar system (i.e., the Sun in our case). As it happens, our Sun is about as "typical" as stars get, which makes it (relatively) boring. Boring is good for us—exciting places in the universe may not be the best for the purposes of life forming and evolving. As a rule, the less massive a star is, the longer it will live. Our Sun, being relatively low mass, will have a total lifetime of about 10 billion years, which leaves a lot of time for life to get going before being destroyed when the Sun dies.

The most massive stars (which do *very* exciting things) only live for as few as a million years or so, and during their brief lives they are not very nice to be around. Even given the very fast timescale that life got going on Earth after the Late Heavy Bombardment, a million years is perhaps too short for any form of life to evolve, and almost certainly far too short for that life to evolve beyond some microscopic state. On the flip side, the lowest mass stars basically live forever (OK, not technically, but for all practical purposes). But—and this is a big "but"—because they are so low mass, the "habitable zone" around them in which we might expect a planet to have liquid water is exceedingly small and close to the central star, which brings about other issues—first, the low likelihood of a planet being found within that small region, and secondly, the orbiting planet would be so close to the central star that it would probably be "tidally locked"—which means the same side of the planet would face the central star, possibly making one side too toasty for life and the opposite side too cold. The word "possibly" here is key, though—and there are ways to sneak around this (for example, something like a band of life-suitable conditions circumnavigating the planet on the light-side/dark-side interface).

As far as life goes, I think there are things in the category of "need to have" and another category of "nice to have." I view energy, liquid, and time as fairly hard requirements for life (i.e., needs), but there are bonus features of Earth that are worth keeping in mind that are nice to have. We don't really know how essential these nice-to-have conditions have been for life, but they probably helped life-as-we-know-it, and they certainly make Earth habitable for humans.

For example, Earth's atmosphere has been vital for life on the surface (although subterranean life might not care as much). For one thing, having an atmosphere is great for avoiding the rapid vaporization of water. Our atmosphere also does a bang-up job of blocking the high-energy X-ray and gamma ray light, large doses of which are lethal to humans. To be sure, some of the extremophiles we talked about could really care less about radiation, so this is admittedly a human-centric view.

The fact that Earth still has an atmosphere is in part due to Earth still having a decent magnetic field. Without a decent magnetic field, the surface of the Earth would be bombarded with charged particles from the Sun (known as the "solar wind"), and that would cause all kinds of problems. For example, without Earth's magnetic field, the solar wind would strip away the ozone layer, exposing life on the surface to nasty radiation. We can look to our sister planet Mars for a lesson learned: Mars was formed with a nice thick atmosphere, but after it lost its magnetic field, that atmosphere was swept away, leaving the dry and barren planet we know and love.

Earth also has a nice massive moon. Not only is the Moon lovely to look at, which surely primordial life deeply appreciated, but its gravity has been helpful in at least two ways. First, the mass of the Moon helps to stabilize Earth's tilt. With a relatively stable tilt, life in different areas of Earth has time to adapt to the local conditions. If the tilt of the Earth were allowed to wobble around on rapid time-scales, there would be drastic climate changes, making it hard for life to evolve and adapt fast enough. Second, the gravity of the moon causes tides, and tide pools *may have* been important for life—in these tide pools the water can evaporate at low tide, leaving highly saturated solutions of organic molecules, making complex combinations more likely to happen.

Then there is the carbon dioxide (CO_2) cycle on Earth that is enabled by plate tectonics. This simple little molecule can and does have an outsized impact on Earth's climate. In the present day, we seem to be doing our best to break this regulation cycle that has been in place for eons. Until the last century or so, CO_2 helped to maintain relatively temperate environmental conditions. When this cycle is working, CO_2 acts like a rubber band pulling back when the temperature swings too far in either direction. This cycle has done an admirable job keeping the climate relatively stable, which has been key for helping life to evolve and adapt. But of course, after millions of years of working pretty well, this one life-form evolved to become "intelligent" enough to have an industrial revolution and throw the whole system out of whack.

The upshot of all of this is that it's hard to get around the fact that Earth *is* pretty darn well suited to life-as-we-know-it. Which kind of makes sense, because if it weren't well suited, life-as-we-know-it wouldn't be here or would be very different (i.e., not "as we know it").

But even if the conditions are well suited for life to survive, it doesn't explain how life emerged to begin with.

How Did Life Emerge to Begin With?

If we want to know if there might be other life in the universe, we need to understand the processes by which it could originate. The general scientific hypothesis is that complex molecules come in contact with each other over time amid different physical conditions, and they are able to form increasingly complex configurations through chemical reactions. Even if the likelihood of any particular chemical reaction is low, given enough time and opportunities for interactions, even rare reactions are bound to happen. Eventually, we end up with things like proteins and amino acids.

Given that we conjecture only a few ingredients are necessary to get life of some form going, you might be wondering whether we've ever tried to just do it ourselves and see what happens. The short answer is yes, we have tried, and no, we haven't succeeded. Yet. But the short answer doesn't quite do justice to the complexity of the topic. There is, of course, a fancy-sounding name for this type of experiment—"abiogenesis," which refers to the processes from which life could have arisen.

One of the earliest experiments to try to identify a chemical origin of life on Earth was started back in the 1950s by Stanley Miller and Harold Urey (and not coincidentally referred to as the "Miller-Urey experiment"). In a nutshell, they took a bunch of chemical compounds believed to be in the atmosphere of the prebiotic Earth, put them in a sealed flask, and shocked them with fake lightning. They expected to create perhaps a dozen or so amino acids, which are the building blocks of peptides, which are the building blocks of proteins, which are the building blocks of life-as-we-know-it. In the 2000s a group of scientists opened a set of sealed vials that remained from the original experiment and found twenty-two different amino acids.

Since the 1950s we've learned a lot more about potential chemistry and the prebiotic Earth. More recent work suggests that clay surfaces could have an important role in helping these lonely amino acids join hands and make nice happy peptides. Now there is a whole new much more sophisticated version of the Miller-Urey experiment in Canada, called a "planet simulator" (which looks amazingly like a very high-tech pizza oven). Maybe by the time this book is published, they will have some early results and this section will already be outdated.

One of the primary hurdles for these experiments, and the researchers working on them, is simple statistics. The prebiotic Earth had a lot more real estate to work with than a few dozen small vials or a pizza-oven-sized experiment. It also had a lot more time to sit around and let things happen than, say, a human lifetime—like thousands of times more time. It may be overly optimistic to expect it to happen on timescales that keep people's attention.

There are also ongoing experiments *that start with* proteins and lipids to see what happens, and whether they can create synthetic cell-like structures. These experiments skip over the part that Miller and Urey were trying to mimic, jumping straight to working with more advanced molecular building blocks. If one experiment can generate the peptides and proteins, and another experiment can take peptides and proteins and generate living cells, then it is only a matter of time before we string them together. Then what?

There is, however, a big fat chasm between the "can we" and "should we" questions. It is easy to get caught up in how cool it would be to see living cells form out of basic materials and everything that might teach us about our origins. I would argue that we should consider taking a step back from time to time and genuinely question whether we *should*—and consider the conditions under which we *should not*. The ethics here are complex, but whenever we talk about doing anything with life, we should really give deep thought to the issues involved. I would rather see our ethical understanding ahead of our science than the other way around. The same goes for terraforming Mars.

One interesting hypothesis in play is that life didn't originate on Earth at all, but rather hitched a ride here from somewhere else. This idea is known as "panspermia." The idea is that somehow (whether by accident or intention) microscopic life traveled through interplanetary space by hitching a ride on a comet, asteroid, meteor, alien spacecraft, and so forth, landed on Earth, and flourished. In this scenario, our *very* distant ancestors were aliens.

Given what we know about extremophiles in our own biosphere, the idea that tiny life-forms could survive interplanetary travel intact is not as crazy as it might initially seem. If life finds a way to withstand the harsh radiation of space (check!), survive ultracold temperatures (check!), and can go into some sort of cryptobiotic state for long periods of time (check!), it has a fighting chance to make an interplanetary journey. Given enough time, this almost seems inevitable, like migratory birds dispersing seeds to distant lands after crossing oceans.

Planets in the solar system have been exchanging meteorites pretty much since the beginning of the solar system. For example, we have identified over two hundred Martian meteorites on Earth—these rocks were ejected from Mars, perhaps as the result of an asteroid impact or volcano, meandered through interplanetary space for maybe millions of years, and through pure chance ended up landing on Earth (a few have likely landed on other planets, too).

Given all of this, I would argue that it is *possible* that the panspermia conjecture is correct. I'm not going to go remotely so far as to argue it is *likely*, but we certainly can't rule it out. At least not yet.

Before we move on from the mystery of the origin of life, I want to mention a disconcerting idea that stretches our notion of life and how it might have arisen. This particular idea also does a great job of reminding us how provincially we tend to think about life and intelligence. In a nutshell, the idea goes as follows: given enough time, it is statistically more likely for particles in the universe to randomly come together and spontaneously form a "brain" (not a literal brain, but an arrangement of particles with similar functionality) than it is for the universe to go through all the steps of making galaxies, stellar systems, planets, life, and finally . . . actual physical brains. In other words, the spontaneous "brains" are a smaller deviation from equilibrium than the requirements to form actual physical brains, which would make them more abundant. The statistical arguments for (and against) these brain-like structures get deep into cosmological predictions about the nature of the universe, and you can be sure that they are hotly debated.

These spontaneous "brains" are referred to as "Boltzmann brains," after the late physicist who helped to invent thermodynamics.[2] Boltzmann brains could even believe themselves to be self-aware and think that they have memories—simply because the arrangement of particles would mimic that perception. Many a physicist has argued that they are certain they are not a Boltzmann brain, but the evidence for this claim typically involves stating that the idea of Boltzmann brains is ridiculous. It would be good to have slightly more logic to back this up than our own misgivings. The upshot is that in a very large (possibly infinite) universe with enough time on its hands, radically different forms of intelligence could arise via pathways far different from the chemical reactions of complex molecules that we hypothesize happened on Earth—whether in structures like Boltzmann brains, or something entirely different.

What of Life Elsewhere?

Boltzmann brains aside, for all these reasons discussed earlier, folks interested in possible life elsewhere in the galaxy (and universe in general) tend to focus their efforts on planetary systems around stars not terribly dissimilar from our Sun. Is that narrow-minded? Probably. Can we envision scenarios in which life might emerge around other kinds of stars? Absolutely. But when we have limited resources and an existing data sample of one planet known to have life, we have to start somewhere, and it makes sense to start where we think success is most likely.

There is one other catch that knocks out half of the candidate stellar systems: most stars are in what are called "binary" systems—meaning they have a partner star and the two stars orbit a common center of mass. This may sound innocuous, and it may be fun for the stars to have a friend to spend billions of years with, but it is not good news for any planets around. In such a binary system, it is virtually impossible for planets to have nice stable orbits. If planets don't have nice stable orbits, the physical conditions on the planet will change radically on short timescales. Let's say some microbial life emerged in a lovely warm tide pool and was adapting well to that environment. If the planet abruptly goes from warm and humid to scorching hot and dry, or frigid and frozen solid, that life doesn't have much hope of adapting fast enough.

We've already seen that just in our solar system there are a couple good places to look for life, so you might be wondering how many more potentially habitable places there might be in the whole galaxy, or even in the entire universe. We have an actual equation for this, which is pretty famous as far as equations go, and named after the late scientist Frank Drake, who was one of the pioneers in the search for extraterrestrial life. The Drake equation works by multiplying likelihoods, and despite its length, it is straightforward.

Be warned: just because we have an equation doesn't mean we have the "correct" answer to the equation. My students often get frustrated with this equation because it doesn't produce a nice tidy confident result. However, the power of this equation is its ability to lay out the variables at play and help us see the impact of different assumptions we might make about them.

$$N = R^* \times f_p \times n_e \times f_l \times f_i \times f_c \times L$$

In the particular incarnation of the equation used here, we are really asking whether we are likely to hear from any advanced ET life in our galaxy. But we can think of this equation like a Swiss Army knife; we can also dial the equation back a bit and ask how much life—*of any form*—is likely to exist regardless of its intelligence and ability to have interstellar communication. All we have to do is drop the last three terms in the equation. Or we can ask about "intelligent" life, *regardless of whether it is communicating*, in which case we drop the last two terms. Or we can zoom out from our galaxy and ask what these estimates are for the observable universe, in which case we get to multiply by the number of galaxies we can see (which is a really big number, so this is fun to do). The point is, there is a lot of flexibility built in, depending on exactly how we formulate the question we want to ask.

In the way this equation is usually presented (as it is here), our confidence in the estimates for the variables gets increasingly lower as we go from left to right. Some of the variables we have a decent handle on (at least by astronomical standards). By the time we get to the end of the equation, the best we can do is speculate. To have any real confidence in the values we adopt for the last couple variables in this equation, I would argue we would need a solid grounding in astropsychology and astrosociology, neither of which exist yet (for the record, I am also excited for astrolinguistics to emerge as a field).

Let's go through these variables one at a time, and I will give you my best range of estimates (and guesses) for these values, but I want you to feel empowered to substitute in your own estimates/guesses as well.

N—**Total number of currently broadcasting civilizations.** This is the holy grail of the equation. In the equation's formulation in this case, N is our estimate for how many technologically advanced civilizations in the Milky Way are currently sending out signals (in other words, with whom we have some hope of communication).

R^*—**The rate at which Sun-like stars form in the Milky Way.** If we assume that life can only form on planets that orbit stars, we need to know how many stars there are. There are a couple caveats, though. First, you will note that this number is a *rate*, which means how many stars are formed per unit of time. We need a *rate* because we need to consider time in this calculation—stars don't live forever, and (probably?) neither do civilizations.

If we only care about Sun-like stars, on average, roughly three of these stars are formed each year in the Milky Way. There is a final caveat, though—at least half of the stars formed are in binary systems, and for reasons we discussed in the previous section, we think those stars are probably not a great stable home for life to grow up in. Taking these estimates, R^* is in the range of one to two stars/year.

f_p—**Fraction of these suitable stars that have planets of any kind.** Assuming that life needs a planet to form on (which might not be true if we are thinking out of the box), we need an estimate of what fraction of the stars (estimated for R^*) have planets. When the Drake equation was originally rolled out, we had no data on what this value might be. In the intervening years we have had major technological advancements enabling us to go out and determine what an increasingly large number of other stellar systems are like. The bottom line is that f_p seems to be 100 percent, or at least close enough to 100 percent to not matter for our purposes here. That makes this variable easy to deal with, because we just put in $f_p = 1$.

n_e—**Fraction of these planets that have a habitable environment.** As promised, moving toward the right in this equation, things start to get a tiny bit more uncertain, but they are not totally un-estimable yet. If our solar system is typical, this value is for sure at least one (since we are here), and may be even as high as three to four if we consider other planets and moons that have (or have had) water. However, when I consider the likely values that n_e might have, I find I need to leave room for stellar systems that just really are not set up for life. If I'm feeling especially conservative, I think this value could potentially be as low as perhaps one-tenth of possible places suitable for life to emerge per planetary system, leaving us with a range of $n_e = 0.1$ to 4.

f_l—**Fraction of habitable planets on which any kind of life emerges.** Of all these planets that might be suitable for life to emerge, on what fraction does life *actually* emerge, even if only briefly? We are now getting out of our nice scientifically testable comfort zone. As of today, we know of one planet in a single stellar system on which life has emerged, and you are living on it. Having a single data point makes it dicey to extrapolate, and we are cornered into having to guess whether we are particularly special. For the moment, let's focus on what we know about life on Earth: as a reminder, we find some type of life everywhere we find liquid water, and life emerged on Earth as soon as it possibly could have. Taken together, these two facts might suggest that life is likely to emerge relatively quickly (on astronomical timescales) on any of these habitable planets that have liquid water.

To be clear, there are a lot of "ifs" involved. And we should be wary of extrapolating too much from a sample of one, which you do not need a degree in statistics to appreciate. On one hand, the emergence of life on Earth could be entirely unique, but pretty much every time in history we've thought that we have some privileged place in the universe, we have turned out to be wrong. Because of this, most scientists are inclined to assume, until evidence suggests otherwise, that we are "typical." In the absence of evidence, a default philosophical position that we are prone to adopting in science is the "principle of mediocrity," which says there is not likely to be anything particularly special or unique about us or our universe. If you go out to a clover field and pick a clover at random, it is unlikely to have four leaves. That being said, four-leaf clovers do exist.

If the principle of mediocrity holds, then f_l should more or less equal 1. This may seem odd, but f_l could also be greater than 1—if life emerged independently multiple times,[3] which is like buying more lottery tickets. However, if Earth is, in fact, a special four-leaf clover, then the question is *how special*? I'm not comfortable with values more pessimistic than about 1/100, but if you want to pick 1/1,000, or 1/1,000,000, go right ahead because I can't refute you with any hard data.

f_i—**Fraction of life that evolves to become intelligent.** Continuing into variables that we have no empirical way to constrain at the moment, what fraction of this life that emerges becomes "intelligent"? And how do we even define "intelligence" anyway? I mean, sometimes we humans are phenomenally stupid. In terms of evolution, this issue revolves around the extent to which intelligence has survival value. There are examples that both support this (like humans, maybe) and refute this (like cockroaches).

For the practical purposes of what we want this version of the Drake equation to do, the definition of "intelligence" in this context is the ability to create technology that could potentially broadcast signals across interstellar space. To be clear, defining intelligence in this way gives me a bit of heartburn because I think it is *far* too narrow of a definition, but in this version of the equation we need to know whether a species could communicate if they were so inclined, otherwise we are unlikely to hear from them.

Even determining what f_i is on Earth is a little tricky. For sure f_i is at least 1 on Earth, because here we are with our big radio antennas. But it isn't like the evolution of species on Earth has stopped—if we wait around long enough, could chimps or dolphins make this leap in intelligence, too? On the other hand, had evolution taken a very slightly different route, we might still be closer to chimps. The principle of mediocrity suggests we adopt f_i=1, but who knows? 1/100? 1/1,000? It really depends on how pessimistic you're feeling today.

f_c—**Fraction of intelligent life that broadcasts technological signals.** If you thought estimating f_i was bad, hold on to your hat. Of the life that becomes "intelligent" (f_i), what fraction goes on to emit signals into space? At least f_i had to do with biology; f_c brings us into the realm of alien psychology, sociology, and politics. In our anthropocentric view, it is very tempting to just assume that *of course* aliens would be curious, and *of course* they would want to know if anyone else was in the universe, and *of course* the alien science foundation would fund projects to do this. We may not quite be ready to make assumptions about what values and priorities ET civilizations might have. Even among cultures on Earth, the adoption and use of technology is deeply heterogeneous. Our good friend the principle of mediocrity suggests f_c is 1 over the course of a civilization's existence. Other than that, you can really pick any value you think is reasonable. Maybe 1/100 feeling appropriately pessimistic?

L—Length of time civilizations emit signals to space. We've made it to L, which I think is the most painful of these variables to consider. In a nutshell, after a civilization develops and uses technology that sends signals into the universe, how long will they be around and do so? This number *really* matters if we have any interest in communicating with any ET life while they are still alive.

Shall we take another quick look in the mirror? We have only had the technological ability to transmit signals to space for about 100 years. Let that sink in for a moment—over the long arc of human history, with our brains being "intelligent" for tens of thousands of years, 100 years is pretty much nothing. Yet during that 100 years we have already developed the ability to annihilate ourselves (and one could argue have come precariously close), which calls into question our working definition of "intelligence" entirely. We are currently in our "technological adolescence," meaning we are "advanced" enough to kill ourselves off, but are we "intelligent" enough not to? Think of it as a civilization not yet having a fully developed prefrontal cortex to help us make good decisions. So, L could be as low as, say, 100 years, but I'm going to go out on a limb and express some cautious optimism that we will make some colossal mistakes, but we will learn our lessons and rebound, which is part of growing up (at least this is the parenting philosophy I've tried to take with my own teenagers). Given the timescale over which humans have been "intelligent," I think 10,000 years is not a totally crazy number to adopt for L.

If "intelligent" civilizations can make it out of their adolescence, then the options open up dramatically. If a stellar system is like ours, their planet might be due for a catastrophic asteroid impact roughly every 10 million years, on average. But I tend to think that if a civilization were even a couple generations more advanced than we are, deflecting an asteroid might not be a big deal. The next key time frame is the evolution of the host star itself—once their star reaches the end of its life, the life on that planet needs to move on, develop some great new technology to harness other forms of energy, or die. This timescale is perhaps a few billion years or so, depending on the host star. If the civilization survives the death of the host star, we are in a whole new territory.

I feel impelled to confess that I am taking several shortcuts here, largely because I'm assuming that someone who picks up this book is not necessarily interested in all the painstaking (and sometimes deeply annoying) details involved. But as they say, the devil likes to take up residence in the details. Modern exoplanet scientists and astrobiologists have modified ways of thinking about the Drake equation, which you may come across upon deeper reading. For example, a new fashion is to refer to a variable called $\eta_\oplus$ (pronounced "eta Earth"), which combines a few of the separate variables in the Drake equation and tells us the mean number of rocky planets that are roughly the size of the Earth and in a potential habitable zone of the host star. There is a whole army of researchers working to determine the value of $\eta_\oplus$ at this very moment, so stay tuned for revisions.

Now that you have a sense of these seven variables and the values we might assign to them, we can put them all together to estimate the number of technologically advanced civilizations sending signals into space at this very moment. Heck, go ahead and put your preferred numbers into a calculator and see what you get. Depending on how optimistic or pessimistic I'm feeling with assumptions about these variables, I find that the chances of there currently being another advanced broadcasting civilization in the Milky Way are in the range of roughly 1 in 10,000 (so unlikely, but not impossible) to there being a billion or more in our galaxy (virtually unavoidable, and they should be everywhere). However, the *only* scenarios in which I find it is unlikely that there is another broadcasting civilization in our galaxy right now are the ones in which I assumed the civilization would survive less than one thousand years after becoming technologically advanced. Sleep well, dear reader.

Of course, we might be interested in *any* form of life, and not just limit ourselves to whether that life ever becomes intelligent and starts to communicate. In this case f_i and f_c don't matter. What do we do about L, though? On one hand, non-"intelligent" life isn't likely to destroy all life on the planet with nuclear warfare. However, it seems to me that nonintelligent life would probably have a hard time deflecting a catastrophic asteroid impact, and is probably even less likely to survive the death of the host star, which would bring us up to timescales of a few billion years or so. I'm also *assuming* that if primitive life starts to emerge, it will find a way to flourish. It could well be that if microscopic life emerges, it will get snuffed out right away by who knows what. But I also think that if the conditions are good for life to emerge, then it will just go on and keep emerging over and over until some form of life gets a foothold in the right place at the right time. For all we know, this is what might have happened on Earth.

If we don't care about whether life becomes "intelligent" or decides to try to communicate, the numbers look a lot more promising, ranging from there "only" being 1,000 (or so) other planets with life in the galaxy to a billion (or so). The primary feasible ways these numbers could be significantly off is if we are either radically misestimating the fraction of habitable planets on which life appears or lots of the life that emerges just doesn't survive for very long, which could be for a whole host of reasons, but probably not nuclear war, because they are not "intelligent."

We do have one more option to consider with this equation. Up until now, we have only been considering the possibility of life in *our own galaxy*. There is a good reason for this: if we are interested in whether we have any chance of communicating with them, even if it takes generations between messages, this hope all but disappears if we start looking outside the Milky Way (barring, of course, some super-advanced technology that might allow cool things like wormholes that could act like subways connecting regions of the universe). But the universe is a lot bigger than our galaxy. If we replace "galaxy" with "observable universe" we get to add a new variable to the Drake equation—the number of galaxies in the visible universe. This is a really big number—around 2 trillion, which is 2,000,000,000,000 or 2×10^{12}. By these estimates, the chances of there being some kind of life somewhere in the universe are, well, astronomically high.

Before we move on, I want to pause here and point out something utterly mind-blowing. The Milky Way is about 10 billion years old.

If life has popped up randomly over this time, and say 1,000 civilizations have come about, this means that, on average, a new civilization pops up roughly every 10 million years. In other words, the next youngest civilization to us—so our closest older sibling—would be something like *10 million years more advanced.* Take a minute to let that sink in. In this scenario, we are—by far—the babies. If we survive, think of what we might achieve in 10 million years. The "if we survive" issue is where things get real, which brings us to a thought experiment that I think is woefully underappreciated.

The general scenario in this thought experiment may well sound like science fiction, but I would argue that even from our current Earthly perspective this is not too far in the distant future. So, if you wouldn't mind, indulge me for a few pages and see how this plays out:

Somewhere and somewhen in the galaxy, life emerged, became "intelligent," and grew increasingly technologically advanced as they endeavored to understand how things work in the universe. Sound familiar? For a time, perhaps things went well with them, but after some period of their existence together on their home planet, something happens. It could be social strife—some type of civil war or identity-based persecution. Or they might run out of resources—land, water, food, and so forth. Possibly their home planet has a catastrophic event—asteroid impact, stellar flare, nearby supernova, and so on (there are a lot of ways the universe can kill life off). Or maybe they just get curious. Versions of all of these have played out on Earth, so hopefully they don't seem far-fetched, because so far we are firmly grounded in the reality of what humans have done. In any case, *something* happens, and a subset of this species decides to try their fates on a different planet, so they embark on a migration to a new home.

I know what (at least some of) you are thinking: *Wouldn't travel times be insane?* Or *What about lifetimes? Wouldn't this type of migration take generations?* All excellent points that could very well be correct. But here's the thing—we are technological babies. Only about 100 years ago we developed the first airplanes, and in that sliver of intervening time we have already developed the technology to go to space, and (as of this writing) the fastest spacecraft so far has traveled at an astounding 1/1,852 the speed of light. I think that it is fair to assume that the coming generations will make technological advances that will blow our twenty-first-century minds—imagine what our descendants might achieve in 1,000 years. I don't think it is unreasonable to predict that coming generations will find ways to reach speeds of 1/1,000 the speed of light.

What about lifetimes though? Even at a speed of 1/1,000 the speed of light, it would take over 4,000 years to reach even the nearest star. Having gone on long road trips with my own family, I don't think I would sign up for this. Can you imagine all of the "Are we there yet?" on the trip? On the other hand, think about the medical advances we have made in the last century. For example, we only discovered penicillin about a century ago, and consider how far the medical field and our understanding of human biology has come since then.

We know of numerous so-called "extremophiles" that can go into suspended animation for indefinitely long periods of time, and cryopreservation is also already a thing, with many labs working to understand how organic cells can be frozen and thawed without damage. Hundreds of (dead) people are already frozen in cryopreservation facilities. It isn't far-fetched to think that in even 100 years we might develop the routine ability to go into states of suspended animation—go to sleep for a few-hundred-year nap, wake up on a new planet.

Also note that our biological evolution has not stopped, and we are going to continue to evolve in some way. For a little bit of context, as a species, we've only been bipedal (walking on two feet) for about 4 million years. In fact, the evolutionary paths of humans and chimps only diverged roughly 4–6 million years ago. How might our brains and physiology evolve in the next million or 10 million years? If we are around that long.

To be sure, travel to other star systems is in the realm of science fiction today, but I don't think it will be outlandish for very many more generations. In which case, let's suspend our disbelief for a bit and consider what our species might do as the above scenario plays out:

1. Intelligent and technologically advanced life inhabits and colonizes a planet.
2. A need arises for some subset of species to migrate to a new planet.
3. They undergo a planetary migration.
4. Go back to the first step and repeat—all the while becoming increasingly technologically advanced.

How long would one iteration of this process take? Once the life form is technologically advanced, which is our starting point, what amount of time might pass before a subset needs or wants to migrate? Looking at human history as example,[4] I think about the perilous journeys that so many people have made due to war, famine, and religious persecution. With history as a guide, I think estimating an average time of, say, 1,000 years for some members of a species to want to head to a new home is not unrealistically short. Already today there are private companies with aspirations to colonize Mars.

The journey itself is the next step. As a nice round number, we can estimate that (on average) stellar systems with habitable planets are sprinkled around the galaxy every 10 light-years or so. Traveling at 1/1,000 the speed of light, this journey would take them 10,000 years. With these (admittedly crude) estimates, this entire process of settling a new planet and embarking on a new migration might take 11,000 years, give or take. Feel free to dial that up or down a bit; it won't really change the conclusions. I also expect that as a species becomes more technologically advanced, this process will accelerate (maybe akin to Moore's law), but I'm not even going to factor that in here because I want to aim for a conservative baseline estimate.

Starting from just a *single* advanced life-form somewhere in the Milky Way, using the estimate above, the process to colonize the entire galaxy might take roughly 10 million years. To be sure, 10 million years does seem like a long time compared to, say, a human life. But to help give a bit of context, 10 million years ago on Earth the dinosaurs had already undergone mass extinction—like 200 million years before that (I won't say they totally went extinct, because I can already hear the complaints of bird watchers reminding me that birds are just evolved dinosaurs). Even in the context of life on Earth, 10 million years is not (relatively) that long. On a galactic scale, 10 million years is even less significant—we can think of the "modern" Milky Way as being about 10 billion years old.

Here is the upshot of this entire thought experiment: If even a *single* species of advanced life arose somewhere in the galaxy, survived, and roughly followed this migration process, *it could have colonized the galaxy more than 1,000 times over.* Please let that sink in. And while we're at it, let's also remember that they would most likely be at least millions of years more technologically advanced than we are.

So where are they?

The Search for ET Life

Finding extraterrestrial life, no matter how primitive, would be one of the most extraordinary discoveries of humankind. Ever. So, of course we are looking. The catch is that detecting primitive life by passively observing adds an extra challenge to an already staggeringly difficult endeavor. As a result, much of our efforts to detect ET life require that life to be technologically advanced enough to send out signals that we could detect over in our corner of the galaxy. What do we look for and why?

A big part of looking for signs of ET life is narrowing down where and how to search based on what we think is likely to be most effective; we simply don't have the resources to look as thoroughly or deeply as we might wish, so we are forced to make assumptions about what ET life might be likely to do. This starts getting precariously into the realm of astropsychology or astrosociology, which are . . . immature fields (to put it lightly). There are countless ways in which the assumptions we must strategically make could be wrong, and there is a mountain of stuff we *know* we don't know. This is almost enough to stop us in our tracks. Almost.

One of the most fundamental assumptions that is typically made in the search for ET life is that *if* they are sufficiently technologically advanced, they will know at least as much about the universe as we do, so our primary method is to look for "technosignatures." One can rightfully question what we choose to define as "technologically advanced," which is a moving target, as each human generation makes new breakthroughs. In fact, even a modestly more technologically advanced alien civilization, perhaps only a few thousand years more sophisticated than we are, may well think of our state of advancement as infantile. Here we simply adopt a practical definition: they have to be technologically advanced enough to communicate across interstellar distances if we hope to detect them across interstellar distances.

Given these assumptions, we extrapolate that ET life would know that one of the easiest ways to send a signal across vast distances in our galaxy would be with radio light.[5] What makes radio light so special? Of all the types of light (gamma ray, X-ray, ultraviolet, optical, infrared, millimeter, radio), radio light has the longest wavelength. For comparison, while X-ray light has wavelengths of roughly the size of an atom, radio light has wavelengths on the order of the size of a human and longer. In fact, there is a whole subcategory of radio light called "extremely low frequency" (ELF, which is a rare case of a great acronym in astrophysics) that includes wavelengths longer than 10,000 km that are even used for practical things like underwater communication with submarines. Radio light can penetrate through material in ways other wavelengths cannot. You already knew this—you can get radio signals in your car or house, but you can't see through your car's hood or living room wall with optical light.

However, there is a catch—isn't there always? Lots of things in the universe have produced light at radio wavelengths too, including the cosmic background radiation (which you might recall from the last chapter is a relic from shortly after the Big Bang) and background light from our own galaxy. This means that for a big swath of radio wavelengths it would be extremely challenging, if not impossible, to detect a faint signal from another civilization above the natural background "noise." Perhaps counterintuitively, this turns out to kind of be a good thing—precisely because most of the spectrum of radio light would be swamped by background radio light, it *narrows down* the region in the radio spectrum that makes the most sense to scan for signals from ET life, making this whole endeavor a tad easier. If you are curious, the wavelength range we believe is optimal for interstellar transmission is from roughly 3 cm to 30 cm, which has been dubbed "the cosmic watering hole."

We have a major added bonus in the wavelength range of 3 cm to 30 cm, but it will take a hot second to explain. One of the things we know about the universe, which we assume another technologically advanced civilization would also know, is this: hydrogen is by far the most abundant element in the universe. Hydrogen also has something called a "spin-flip transition" that corresponds to a wavelength of 21 cm, right smack in the middle of the optimal wavelength range for sending signals across the galaxy. This 21-cm spin-flip transition is one of the most important wavelengths for understanding the universe,[6] and it would stand to reason that if someone else out there is trying to understand the universe, they might be using that wavelength, too. And, if they make the same assumptions that we do, they would assume that we would assume that they would try to communicate at or near that wavelength. I know, that is a heck of a lot of assuming.

Thus, our line of reasoning goes like this: We expect a technologically advanced civilization would know at least as much about the universe as we do. Therefore, they should know about the "cosmic watering hole" (though, admittedly, probably by a different name). They should also know about the importance of hydrogen, and they should know about the 21-cm spin-flip transition, which pinpoints a range of wavelength of light to monitor. So that's what we do, mostly.

Radio light has been the workhorse in searching for ET life pretty much since the beginning of searching for ET life, but that doesn't mean we are turning a blind eye to other options. For example, it is possible that ET life might use something like lasers. Lasers are great for long-distance signaling because they have highly focused beams that minimize the amount of energy needed to transmit a signal. Another major bonus of lasers at optical wavelengths is that a lot of information can be packed into the signal. In fact, we humans are already using lasers to communicate with satellites in space, so the foundation of this technology is already available here on Earth.

The catch (remember, there always seems to be a catch) is that for us to notice a laser signal from an ET civilization inhabiting a planet orbiting a star, the brightness of the laser beam would need to be significant enough to stand out from the star their planet orbits. We don't have the technical ability (yet) to resolve planets from stars at interstellar distances, so the light of the star + laser would have to be sufficiently brighter than the light of the star alone. Exactly how bright the laser flare would need to be depends on the star the ET life home planet orbits, but even with our relatively adolescent technology on Earth, we are on the verge of being able to send a signal like this to the closest star systems. If we wanted to. So, we assume that advanced ET life is likely to have this ability, too (if they wanted to). There are, of course, reasons we and they might not want to.

What about nontechnologically advanced life? The overwhelming majority of life that has evolved on Earth is not technologically advanced and may never be. Given those statistics, it seems likely that nontechnologically advanced life in the universe vastly outnumbers technologically advanced life.[7]

If there are alien planets teeming with primitive life, detecting it seems extra challenging. Never fear, we have learned how to take scraps of data we get from the universe via light and decode all kinds of information. While we are a long way from having resolved images of planetary surfaces in other star systems—like satellite images we have of Earth showing cars, streets, buildings, trees, and the occasional blurred-out person—we have another trick up our sleeve. With the most cutting-edge telescopes we have today, we can start to determine what the atmospheres of exoplanets are made of.

Having an atmosphere at all is a good start if we are looking for life-as-we-know-it. But we can do even better than that. With modern observatories we are beginning to be able to identify biosignatures in these atmospheres. There are specific molecular compounds that may require biology (e.g., life) to be generated at high levels. One promising (if stinky) molecular compound is methane. Did you know a single cow can produce over two hundred pounds of methane each year? To be sure, we don't expect there to be actual cows happily grazing on alien planets, but methane is a general waste product from microbial metabolisms. There are lots of other compounds that are promising as well—sulfur gases, nitrous oxide, methyl chloride, and so on, which have strong biological contributions on Earth.

The keys for this detection technique to work are (1) there needs to be sufficient biomass to produce enough of whatever it is we are trying to detect, (2) the specific molecular compound needs to otherwise only exist at low levels due to abiotic production, and (3) it needs to have a signature in the spectrum that we can detect. I don't think we will have to wait long to observe potential biosignatures from other planets, but I do think we will need to try to restrain ourselves from prematurely jumping to conclusions. Remember, extraordinary claims require extraordinary evidence, and we will need to carry out due diligence to rule out other prosaic nonbiological origins. Of course, the "due diligence" part is not usually as flashy or exciting, but it would give us a lot more confidence that we were on the right track.

You may be wondering whether we have found any signs of ET life yet, and you might think this would be a yes or no kind of question. However, reality doesn't always make things so easy. Much of the problem is that observed anomalies have not been kind enough to repeat, which means we don't get the chance to test them and rule out more vanilla explanations. A great example of this is the "Wow! signal" that was observed in 1977.

An extremely strong radio signal was detected by the Big Ear radio telescope (I'm not kidding about the name) in Ohio at almost *exactly* the wavelength of the 21-cm spin-flip transition of hydrogen that we think is so important and right in the "cosmic watering hole." Data visualization has come a long way since 1977, but at the time the value of the signal strength was read out as numbers and letters, with 1–9 being weak–strong, respectively, and then switching to letters A–Z for signals that were extra strong. Imagine that it is your job to read these strip charts every day, and day after day you see a field of blank spaces and mostly 1s, 2s, and a few 3s. Then one day you go in to read the chart and clear as day there is "6 E Q U J 5," which lasted just over a minute of time. You might scribe "Wow!" next to it in the margin as well.

Credit: Big Ear Radio Observatory and North American AstroPhysical Observatory (public domain).

Before you ask, yes, folks have gone back many times over the intervening years to look for signs of the signal repeating, but it hasn't repeated (at least not while we were watching). There was also a conjecture that it could have been due to a comet that we didn't know about in 1977, but that has essentially been ruled out. A whole slew of other natural hypotheses has been floated to explain the signal, including a radio source on Earth reflecting and bouncing back from a chunk of debris in space. Or maybe it was some kind of beamed burst of emission from across the galaxy or distant universe or some kind of interstellar scintillation. Or maybe it was ET life.

I sometimes imagine that the "Wow! signal" was, in fact, from an advanced civilization that was facing its demise, and in one last Hail Mary tried to send us a message of warning. Who knows, maybe that's true. The thing about not being able to test a hypothesis is that we can't rule it out.

We *mostly* passively watch for potential signals from ET life. Thanks to a long list of movies and TV shows that portray ET life as hostile and violent, many people are not super keen on sending out a beacon letting other potential life in the universe know we are here. This brings us back into the sketchy and utterly undeveloped area of alien psychology. There are arguments to be made for advanced ET life potentially being hostile, but there are also arguments to be made supporting a hypothesis that they are likely to be benevolent. I am pretty sure we project a lot of our own human psychology onto our imaginings of the motives and character of ET life. *Of course*, we would assume ET life is out to colonize and consume resources while removing threats—because that is exactly what humans have done. However, it's worth reminding ourselves that ET life need not have evolved with the same psychological characteristics as humans (or any other species on Earth for that matter).

Let's come back to that word "mostly" at the start of the previous paragraph. We have been passively broadcasting signals into space for a very long time. Every radio broadcast, every TV show, every cell phone call, and so forth that makes it past Earth's ionosphere is sending its signal out into the universe, where it is gently bobbing along on electromagnetic waves and gradually getting weaker and weaker. Not to alarm you, but the first high-powered radio broadcast was of Hitler and the opening ceremony of the Olympics in Nazi Germany in 1936. I'm personally hoping that any ET life within one hundred light-years doesn't have advanced-enough technology yet to detect that particular signal.

Aside from passive broadcasts, people have *intentionally* sent about a dozen strong signals out into the cosmos, most of them beamed at specific locations. For example, we sent a series of transmissions back toward the direction the "Wow! signal" came from. You will be delighted to know that our reply to the "Wow! signal" included Twitter messages from the public and a message from Stephen Colbert explaining that humans are not good to eat.

Most of the broadcast messages have been targeted at relatively nearby star systems (< 70 light-years away), which would help facilitate a two-way conversation—taking a mere 140 years to possibly get a reply. Definitely not a good graduate student project. The first high-powered message to reach its destination will be "A Message from Earth," which was sent in 2008 and will reach the Gliese 581 star system in 2029. This star system was specifically targeted because it potentially has terrestrial planets orbiting within the habitable zone of the host star. Stay tuned in 2048 for a returned call.

The Fermi Paradox

As the story goes, one summer day the late physicist Enrico Fermi walked to lunch with his colleagues, discussing the possibility of ET life (as one does), and then later over lunch—and out of the blue—Fermi just blurts out, "But where is everybody?" Then he got a whole paradox named after him. Ah, the good old days.

The thing that was bothering Fermi, and I hope now will bother you, is this: given some basic assumptions about life, the galaxy really ought to be teeming with it. So where is it? Or alternately, where have our assumptions gone completely wrong? There are four broad categories of solutions to Fermi's paradox, each of which has profound implications for life in the universe and the future of humanity.

Category 1: Other intelligent life just hasn't emerged. Perhaps life emerging on Earth just about as soon as it possibly could have and "intelligence" having some survival value are flukes, and we are the best the universe has to offer.

Category 2: ET life does exist out there somewhere, but we just haven't seen it. Maybe we just haven't looked hard enough or in the right way. Or it could be that the ET life is out there but has no interest in communicating. I mean, have you seen what humans are doing? Or ET life is out there, but they are millions of years more advanced; they might know we are here and not care, or they might care but be waiting for us to grow up. I feel a fair amount of confidence that if they are millions of years more advanced than we are, and they don't want to be seen, we are not going to see them.

Category 3: ET life exists, and it has been seen, but it hasn't been acknowledged or there has been a cover-up. This is an evergreen favorite, and I am certainly not naive enough to say that our governments don't hide things from us. But having spent many decades among scientists, we are just not that good at acting. A signal from ET life would spread like wildfire, especially as we scrambled to get every telescope on (and off) Earth on the signal source. I know, this is exactly what someone in on the conspiracy would say, so I guess you'll have to decide whether or not to believe me. Could ET life have contacted governments and bypassed scientists? I guess it is possible, but I'm also not sure that the magnitude of this secret could be kept by the number of humans who would necessarily know about it. But I suppose we can't conclusively rule any of this out, either.

Category 4: There has been ET life, but they have died off. This is by far the most depressing and alarming scenario. What if all the assumptions and estimates we've made so far are not too far off, and life has developed thousands or millions of times throughout the galaxy. BUT . . . before it reaches the point that it can migrate it gets destroyed or destroys itself. As we've seen, there are a lot of ways the universe can kill us, and that sucks, but so be it. What I'm not at all OK with is how easy it would be for us to bring about this destruction ourselves. My real fear, though, is that when civilizations reach their technological adolescence, they are not "wise" enough to avoid destroying themselves. We have only been technologically advanced for about a hundred years. "Mutually assured destruction" was the phrase during the Cold War.

You are, of course, welcome to consider which solution to Fermi's paradox you find most likely. My own suspicion is that reality sits in either Category 2 or 4, but I really hope that it is Category 2. If Category 4 is correct, that is a big warning to us that our existence on this planet is much more fragile than we would like to admit. Not only is whether or not ET life exists an unsolved mystery of enormous magnitude, the potential answers may also have profound implications for humanity itself.

ET Life and Beliefs

I often think about how discovering extraterrestrial life would actually affect us humans. Certainly, there would be news stories, magazine covers, and probably a Nobel Prize, I'm guessing a marked uptick in the number of kids interested in science, and perhaps some extra funding. But what I really want to know is how the discovery would impact the deep psychology and beliefs of any random person in their kitchen making breakfast. To my mind, this would be in the running for the most important discovery in all of human history, but on my cynical days (which seem to be an ever-growing fraction of them), I have a nagging fear that there would be a flashy news cycle and then most folks would just go back to their everyday lives, making toast and coffee and just not thinking about it.

To be sure, how people might react to a confirmed discovery of ET life will almost certainly depend on the form that life takes. If we find some microscopic life-form, barely more than a collection of complex molecules, maybe the collective population will just kind of shrug. On the other extreme, if an advanced civilization made contact with us, I feel confident that many people would panic. You might be relieved (or shocked) to learn that there are a handful of psychologists who are trying to proactively think about how humans would respond to the discovery of ET life. Given that we don't actually have any hard data to work with, it is not surprising that their predictions cover the map,[8] but in fairness, individual human reactions will probably cover the map, too.

Because I grew up in a religious tradition that rebuked belief in ET life, I am also interested in how discovering other life in the universe might impact one's religious or spiritual beliefs, particularly those that are Earth-centric. This is an actual area of study known as "exotheology." Many theologians of world religions will argue that the existence of extraterrestrial life is not inconsistent with their core texts and beliefs, which may well be true. But—just to try my hand at becoming Captain Obvious—what the religious texts actually say and how they are interpreted are often not the same thing. Many religious traditions place humanity at the center of creation and see humans as unique and special handiwork of a higher power. Other religions might even see ET life as evidence of divine creation. Still others might struggle to incorporate this new knowledge into their existing beliefs and may have a crisis of faith.

When I was growing up, we only had one small shelf with books, and at least half of them were different versions of the Bible, so I spent a lot of time reading them. (I note that one of the other handful of books was a Spanish-English dictionary, and my ability to speak Spanish is virtually nonexistent, so perhaps the lesson is that reading books and understanding them are not the same thing.) I recall a conversation I had with my mother while I was in elementary school in which I told her I *wanted* to believe in ET life, but it was against my religion. There is a lot to unpack there.

Regardless of what religious texts actually say, it is clear that many Earth-centric interpretations are latent in our day-to-day beliefs, and perhaps so subtly entwined that we don't even realize how they impact our mental schemas. How would we feel about having a "sibling" in the universe? How would stories like Adam and Eve have played out on other planets? Would aliens even have souls? Or does that depend on how intelligent they are and what they look like? What if ET life is far more intelligent than us? I've come to think that examining our underlying beliefs and how robust or flexible they are with respect to extraterrestrial life has the potential to teach us a great deal about ourselves, whether or not we ever discover ET life.

Interstition

A resolution to any of the mysteries in this book has the potential to alter the course of humanity—both through the new physics that would be unlocked and our perception of ourselves. Answers to most of the mysteries in the coming chapters may be out of our reach—if not forever, at least in the near future. However, before we start a steep uphill climb through the potentially unknowable, there are two other great mysteries that could be solved in our lifetimes: What are dark matter and dark energy?

Earthly historians have an easy job by cosmic standards. To state the obvious, a lot has happened between the Big Bang and now. Somehow, we go from "not nothing" to "life," but in the intervening years there is a lot that we don't (yet) understand. Chief among the not understood ingredients shaping the universe are dark matter and dark energy, which affect how objects move in space and the very properties of space itself. This is especially frustrating because nothing less than the fate of the universe depends on the nature of these dark elements.

The good news is, both dark matter and dark energy can be pursued empirically. The bad news is that this chapter may be out-of-date in the not-too-distant future. It will be fascinating to look back in a few decades and assess how much our understanding has changed.

5

What Are Dark Matter and Dark Energy?

THE NEXT TIME YOU MIGHT BE TEMPTED TO FEEL LIKE SCI-
entists have a good handle on how things in the universe work, just
remind yourself that at the moment we only understand what about
5 percent of the mass-energy in the universe is. Don't get me wrong,
there are still questions about exactly how this 5 percent behaves,
which constitutes a hefty fraction of what astrophysicists study in
their daily lives, but at least we know what this 5 percent is made of—
it includes everything you would think of as "normal" stuff in the
universe: stars, gas, elements in the periodic table, neutrinos, and so
on. The other 95 percent of the mass-energy in the universe falls into
the categories of dark matter and dark energy.

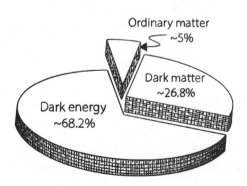

In the cases of dark matter and dark energy, the word "dark" means that whatever this stuff is, we can't see it. More than just not being able to see it, we don't think this stuff interacts with light *at all*; if it doesn't absorb, emit, or reflect electromagnetic radiation, astronomers are left without our primary tool for investigating the universe. What we *can* see is how this dark matter and dark energy interact with other things in the universe, and from that limited information we try to make inferences about what their properties are, make predictions based on those properties, and try to test these predictions. This is arduous work with a lot of null results.

Aside from having the word "dark" in their names, and currently vexing astrophysicists, dark matter and dark energy have virtually nothing in common. Therefore, we will take each in turn. You may well be wondering why we think this dark stuff exists at all if we can't see it, so that is a good place to start.

Dark Matter

Of all the sections in this book, this is the one that I'm betting will need revision on the shortest timescale. While some of the other mysteries in this book may not even be solvable, regardless of how much our technology advances, experiments to figure out the nature of dark matter are proceeding apace. If these experiments could hold off on making that discovery until after the first edition of this book has been out for a bit, I would greatly appreciate it. Even if there is a discovery in the meantime, I hope that this section might help you put the news in context.

Arguing that something exists that we can't see or interact with requires a pretty high bar. There are really two parts to our inference of dark matter that are required to get over the skyscraping standard: First, why do we think there is something out there at all? And second, why do we think this "something" isn't one of the usual things in the universe? We'll start with the first.

Have you ever thought about how we know *anything* about masses of objects in space? After all, it isn't like we can bring them home and put them on a scale. To determine masses of celestial objects, we look at how they move, and how they move is dependent on well-understood gravitational dynamics. Consequently, to understand the discovery of dark matter, we need to invoke a tiny bit of gravitational dynamics, but nothing too crazy. If you had a high school physics class, the following equation for orbital velocity might look familiar.

$$v = \left(\frac{G\,M}{r}\right)^{\frac{1}{2}}$$

As usual in physics, v is the velocity of the thing doing the orbiting, M is the mass around which the thing is orbiting, r is the radius of the orbit, and G is the gravitational constant, which we can ignore for the time being. For our purposes here, what matters is how the variables (v, M, and r) play off each other. For example, if you want v to be larger, you have two choices—either increase M, or decrease r. Let's work through a couple different scenarios to see how this equation maps into behavior we can observe.

Expected rotation curves can (and do) get much more complex (if everything in the universe were a simple curve our jobs would be a lot easier, but the universe would probably be boring), but the underlying physics is straightforward. Using gravitational dynamics, we can harness the relative positions and velocities of all kinds of objects in the universe to infer essential properties like mass, which is the subject at the core of dark matter.

One of the most foundational papers in this area was published by Vera Rubin[1] and Kent Ford in 1970.[2]

Analyzing new (at the time) observations of the Andromeda Galaxy,[3] it was clear there was more going on than meets the eye. Literally. The rotation curve of the Andromeda Galaxy showed that there must be a heck of a lot more mass in the galaxy—particularly at large radii—than could be accounted for by understood components of the galaxy. What's more, when astronomers followed up by measuring the rotation of the Andromeda Galaxy with *invisible* light (particularly tracing hydrogen gas), it was clear that the rotation curve continued to indicate more and more mass in outer regions that we could not account for. The rotation curve just kept climbing.

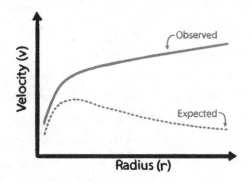

Over the intervening years, veritable mountains of evidence have piled up from observations of gravitational systems across the visible universe, so whatever is going on is pervasive and not just a few anomalous systems. We've also found clear evidence for dark matter using completely independent techniques across disparate observations ranging from gravitational lensing[4] to the cosmic microwave background. We have been forced to conclude that at least one of the following is true: there is a ridiculous amount of mass in the universe that doesn't interact with light, or our understanding of gravity is very wrong.

In the case of dark matter, we have systematically eliminated a range of possibilities, and the list of what remains is growing short. The possibilities that remain include primordial black holes, undiscovered particles, and . . . wait for it . . . other dimensions that share their gravitons with us.

How do we even begin to rule out candidates for dark matter? One major feature of dark matter that we can leverage is that it has a "temperature," which is not an actual temperature (totally on-brand for physics). In this case, what "temperature" refers to is a measure of how antisocial the dark matter candidates are—or rephrased, how far could they go in a straight line at a party and not interact with someone. "Hot" dark matter candidates generally move at close to the speed of light and can travel great distances without even so much as a nod hello. Before you know it, they are on the other side of the room. I will admit to having colleagues who can travel great distances without interacting with anything and remind me of hot dark matter at parties. On the other hand, "cold" dark matter candidates tend to stay where the action is; they may not want to talk to anyone, but they also don't have the motivation to go anywhere else. If you want to get technical, we call the level of social behavior of a particle at a party the "free streaming length"—how far a particle will propagate in the universe before interacting with something. By narrowing down the free streaming lengths (aka "temperatures") that match observations, we can rule out entire categories of dark matter candidates.

Intuitively, you can see how these dark matter temperatures would affect their environments. As with many classic physics demonstrations, let's invoke balloons as an analogy (if any balloon company CEOs are reading this, you should really look into getting astrophysics endorsements—or better yet, supporting astrophysics). If the air inside a balloon is hot, the balloon will be nice and puffed up. If you put that same balloon in the freezer (another experiment you can do at home!), after the air inside cools down, the balloon will look very sad and deflated. After all, there is a reason hot-air balloons are not cold-air balloons. Similarly, if dark matter is "hot," it will be puffed up—big and fluffy. Since the majority of the mass in a system is in the form of dark matter, and all the other "normal" stuff responds to the gravity of the dark matter mass, having hot dark matter results in bigger, puffier size scales of gravitating regions in the universe (think nice yeasty bread). On the other hand, with cold dark matter, gravitating regions tend to be smaller and clumpier (think gnocchi). In addition to hot and cold dark matter, there is also a category for warm dark matter; for the sake of stating the obvious—motivated in part because our terminology doesn't always make sense—warm dark matter is in between hot and cold.

We can test predictions from different-temperature dark matter candidates in a couple independent ways. Both the cosmic microwave background and the formation of galaxy structures in the universe are strongly dependent on how the dark matter behaves, and in both cases the evidence solidly points to dark matter being cold.[5]

The necessity for cold dark matter rules out a whole set of possible candidates in one swipe. Among the candidates ruled out are neutrinos, which were a fan favorite for many years, at least in part because they are known to exist (in stark contrast to some of our other options). To be sure, neutrinos do have mass (albeit incredibly small), and this mass contributes to the underlying mass-scape of the universe, but even if dark matter had been found to be "hot," neutrinos are not nearly abundant enough to account for the mass in dark matter.

The first category of cold dark matter candidates to consider are the so-called "massive compact halo objects" (MACHOs).[6] When dark matter happens to come up in social conversation (as it is prone to do), it is astounding how often one of the first things people ask me is "Have astronomers thought about black holes?" I get where they are coming from, I really do, and I applaud folks for immediately going through the Rolodex of possible options in their brain for something that would make sense. Black holes might nicely fit the bill if you don't know otherwise. But I am also amused by their assumption in the subtext that legions of astrophysicists have been thinking about and working on dark matter for decades and somehow didn't manage to think of black holes as an option.

Black holes are an example of a type of MACHO. MACHO candidates include a host of astronomical objects that are very hard to see—things like neutron stars, rogue planets, and brown dwarfs (which are basically not-quite-stars). The good news is that we can observationally test the abundance of MACHOs. Ripping off the Band-Aid, the bad news is that there aren't remotely enough of these types of objects to account for dark matter. We can look for MACHOs by, once again, harnessing gravitational lensing; when a MACHO passes in front of a distant light-emitting object, the light from that object is bent a teeny tiny amount (we call this microlensing) and gets amplified. Several independent groups have carried out surveys to look for this microlensing—which they do by monitoring millions of stars over many years, from which they can determine the population of MACHOs. Yes, there is an occasional microlensing event, but they are rare, and the result is that MACHOs can't account for more than 8 percent of dark matter.[7]

There is another issue with MACHOs, which is important to note because triangulating conclusions from independent lines of evidence gives us way more confidence that we aren't fooling ourselves. In this case, the problem with MACHOs goes all the way back to the Big Bang, and the primordial nucleosynthesis that came up in the last chapter. To put it bluntly, there is no feasible way that enough baryons (I realize we haven't talked about baryons yet, but we will soon; in the meantime, think of baryons as "ordinary" matter) could have been created during Big Bang nucleosynthesis to account for the amount of dark matter.[8]

Next up are primordial black holes. We just ruled out "normal" black holes, but there is a remaining loophole we can exploit. If small black holes were formed out of high density in fluctuations in the very early universe (i.e., before nucleosynthesis), they just might work. The trick is that they must have been formed especially early in the universe (so they don't take up baryons), be in a mass range that would not have been detected by the MACHO experiments, and also be massive enough to not yet have evaporated from Hawking radiation (more on this in the chapter on black holes). This might sound like it would rule out most possible masses (and it does), but there are still slivers of parameter space in which primordial black holes could have a home.[9]

One lovely thing about this hypothesis is that it relies on physics that we think we know and understand. Perhaps even more importantly, it is at least partially testable. For example, primordial black holes could be detected through gravitational lensing (outside the range that the MACHO experiments have already looked), and through gravitational wave detection when these black holes rotate around each other or merge. Primordial black holes could also explain some existing discrepancies between observations and theory—for example, an apparent overabundance of X-ray light.

For reasons I don't fully understand, this conjecture for dark matter has not seemed as popular in the astrophysics world as it might deserve, perhaps because the available parameter space is relatively confined. That being said, given that this idea is both based on existing physics and is testable, I wouldn't take it out of the running yet.

One facet of dark matter that is imbued with poetic sensibility is the possible connection between the very largest and very smallest scales, including possible undiscovered elementary particles. This option has been a leading contender for decades. This giant umbrella category includes a host of hypothetical particles that may or may not exist. In general, we are looking for any candidates that can meet the following criteria: they must be (1) cold (i.e., not zip around the universe at relativistic speeds); (2) interact with gravity (i.e., have mass); (3) be electrically neutral (or they would interact with light); and (4) be stable and not decay. In creating a list of particles that can meet these criteria, we are forced to consider options outside the standard model of physics, which is both exciting and unsettling.

It turns out that supersymmetry theories in physics (of which there are dozens) naturally give rise to stable particles that would fit the bill for dark matter (dubbed "weakly interacting massive particles," or WIMPs—just to vie with MACHOs on the acronym front). Experiments to detect WIMPS have been going on for decades, reaching lower and lower detection limits. Every year when I teach this material, I expect I will have to revise it before I teach it again. But, at best, the only thing I need to update are the detection limits. To be fair, some tantalizing results have come out of experiments such as the Cryogenic Dark Matter Search (CDMS) that got our hopes up, but to date the evidence has not been compelling.[10]

In addition to WIMPs, other not-yet-observed particles are still feasible dark matter candidates, too. Leading contenders here are axions, which—I kid you not—were named after laundry detergent (the alternate name being "Higglet").[11]

Axions were originally introduced to solve another long-standing problem in physics, which is related to the charge-parity (CP) violation that will come up in the chapter on time. If axions exist, they should pervade the universe. Coupled with their otherwise dark-matter-friendly characteristics, they seem promising. If only we would detect them; as with WIMPs, to date the only thing experiments have been able to do is put limits on their properties—if they exist.[12]

If hypothesizing particles that continue to not be detected and are not known to exist doesn't sound desperate enough, another option for dark matter being considered are other dimensions. These "bonus" dimensions could either be "compactified" or macroscopic (more on this in the chapter on dimensions), and each scenario gives rise to potential sources of dark matter.

For example, in the "universal extra dimensions" model (known as "UED"), the standard familiar forces in our "normal" dimensions can also propagate in compactified dimensions. Because the fields would have to fit within these compactified dimensions, their possible values would be quantized (i.e., have specific discrete values).[13]

In turn, the quantized modes give rise to associated particles (called "Kaluza-Klein particles"), which would be hidden from view in our current experiments; we would need to achieve dramatically higher energies to probe these size scales. However—and this is the key—the momentum (and inferred mass) of these Kaluza-Klein particles, would still be felt in the normal dimensions of the universe. In other words, we end up with invisible particles having a macroscopic gravitational influence.

Macroscopic extra dimensions also create a mechanism for gravity to be felt by unseen mass. In brane-world models (again, see the upcoming chapter on dimensions), which incorporate string theory, gravitons are closed strings that form a loop, which means they are not anchored to a brane in the same way open strings are. Consequently, gravitons can move *outside* of a brane, and potentially be felt by other branes—or even our own brane folded back on itself.[14]

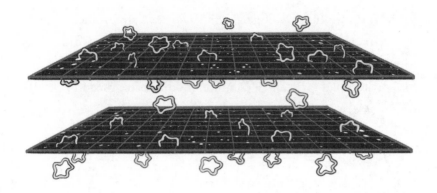

As crazy as this idea might sound, it has been taken seriously by a significant number of physicists, with papers amassing thousands of citations.[15] There have even been multiple experiments in an attempt to place constraints on the existence of parameters of extra dimensions.[16]

For example, scientists are using properties of observed black holes,[17] poring through data from the Large Hadron Collider looking for signatures of extra dimensional gravitons,[18] and even watching for the signature of a micro–black hole being formed.[19]

Finally, what if our understanding of gravity is wrong? Do you remember the "What if?" type of axiom from Chapter 2? It has come back for a visit. It turns out that our empirical understanding of gravity in weak gravitational fields is on shaky ground because it is extraordinarily difficult to test. There are labs on Earth that work on extremely precise measurements of gravity, and the gravitational constant G that keeps showing up in our equations has been measured to a precision of $6.67408 \pm 0.00031 \times 10^{-11} \, \text{m}^3 \text{kg}^{-1} \text{s}^{-2}$.[20]

The problem is that we can't just set up a lab on Earth and dial Earth's gravity up and down to see what happens, so we are stuck trying to measure gravity in a relatively strong regime (at least compared to the distant outskirts of galaxies). Advocates of modified Newtonian dynamics (MOND) hypothesize that gravity behaves differently in weak regimes. If we compare the standard Newtonian equation for gravitational force and a common modification used in MOND side by side, you can see the difference:

$$F_{Newtonian} = \frac{G\, m_1\, m_2}{r^2} \qquad\qquad F_{MOND} = \frac{G\, m_1\, m_2}{\mu\left(\dfrac{a}{a_0}\right) r^2}$$

In the case of the MOND equation, the variable a is the usual acceleration due to gravity, but it is now modified by a new constant a_0 and an "interpolating function" μ. As a result, the MOND formulation enables dynamical behavior that could appear as though an object is subject to a stronger gravitational field than we would expect if we assumed standard Newtonian dynamics.

In my view, MOND is an example of great scientific thinking and not being afraid to question our fundamental axioms—in this case asking if dark matter doesn't exist, how would gravity have to behave to explain the observations? Unfortunately, observational tests appear to rule out the MOND hypothesis, although MOND advocates work tirelessly to find loopholes that might keep MOND afloat.

One of the most damning bits of evidence against MOND involves a pair of massive clusters of galaxies that collided, known as the Bullet Cluster.[21] Clusters of galaxies colliding sounds like a violent event, but an essential feature for the point at hand is that the galaxies in clusters are relatively small and far apart on the scale of the cluster, and the stars in the galaxies themselves are almost point-like with huge spaces between them on the scale of galaxies (do you remember the grapefruits from Chapter 1?). As a result, when galaxy clusters collide, the galaxies generally pass right through each other without much interaction—we call this a "collisionless" interaction. On the other hand, *in between* the galaxies in clusters, space is filled with gas (which we call the "intercluster medium" [ICM]), which typically has roughly 10 times more mass than the galaxies of the cluster themselves—so this intercluster medium is not to be trifled with. When the intercluster media of the two clusters run into each other, they are very much *not* collisionless, instead smacking together and decimating their respective forward motion. As a result, the galaxies (which are collisionless) keep on moving, but the intercluster medium (with is collisional) gets stuck in a pileup in the middle.

Now we layer dark matter and MOND into this scenario, which make different predictions. If dark matter exists, it doesn't seem to interact strongly with anything in the universe other than gravity, which suggests that it is *collisionless*, like the galaxies in the clusters. As a result, when the galaxy clusters collide, the individual galaxies and dark matter will (more or less) pass right through each other. In the dark matter scenario, the vast majority of the mass in the system is in the form of this dark matter, and therefore the overwhelming bulk of the mass will keep on moving in the direction it was originally headed. On the other hand, if there is no dark matter, but instead our understanding of gravity needs to be tweaked, then the majority of mass in the system will stay centered on the intercluster medium, which gets hung up in a traffic jam in the middle of the colliding systems. Thanks to human ingenuity and physics, we can observationally test which of these scenarios has played out; light from galaxies in the distant universe behind these colliding clusters will be distorted (gravitationally lensed) from the mass in the system, and exactly how it is distorted gives us a map of the distribution of mass in the colliding galaxy system. I'm guessing you've already inferred the answer, because we probably

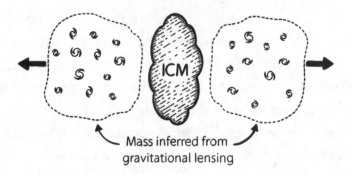

Mass inferred from
gravitational lensing

wouldn't have this chapter on dark matter at all if MOND had been verified.

Advocates of MOND will be quick to point out loopholes through which our understanding of gravity still has room to be modified. But most of the astrophysics community is convinced that— even if our understanding of gravity is tweaked—there must be dark matter.

Taking stock of the dark matter situation, we have a whole busload of options that tap into different aspects of physics, which strikes me as a healthy situation. Curiously, the dark matter candidates that seem the most viable in terms of *known* physics have largely either been ruled out (e.g., MACHOs) or continue to evade experiments (e.g., WIMPs). Other options, like primordial black holes, almost *have to* exist at some level if our understanding of the early universe isn't too far off, and even the most eccentric ideas (like other dimensions) still have credibility (at least to my mind). I actually wonder whether *several* of the dark matter candidates might bear fruit, in which case we might ultimately (and ironically) end up in a situation with *too much* dark matter predicted in the universe. In any case, I stand by my conjecture that this section will need revision sooner than later.

Dark Energy

The discovery of dark energy is a magnificent example of the scientific process unfolding step-by-step and ultimately overturning long-held conventional wisdom. I was a graduate student when the initial observations were published, and it has been fascinating to watch the scientific community (including me) transform from highly skeptical to ubiquitously accepting. As the late Carl Sagan said, extraordinary claims require extraordinary evidence, and this was a case where that standard was overtly deployed. We may generally accept dark energy as existing now given the extraordinary evidence, but that doesn't mean we understand it.

Why do we think dark matter exists to begin with? In the 1990s we'd known for many decades that the universe was expanding, and the quest had transformed from not just confirming the expansion, but determining *precisely* what this expansion rate was. Nailing down this value was a high priority because it not only constrained the age of the universe, but it could also tell us how the universe would end. Debates ensued and arguments were carried out through scientific papers.[22] One of the first key projects for the Hubble Space Telescope was to precisely determine this expansion rate, called the "Hubble constant" (H_0), which has landed on a value of about 70 kilometers per second per megaparsec[23] (km/sec/Mpc). The units of "km/sec" may not seem that odd to you, but the "Mpc" may throw you off: these are units from measuring the rate of change of length (e.g., velocity in km/s) over a specific distance (Mpc).

$$v = H_0 D \quad \rightarrow \quad H_0 = \frac{v}{D}$$

Our primary means of determining the Hubble constant is by looking at the relationship between how far away a galaxy is and how fast it is moving away.[24] Determining the velocity of a distant galaxy is "easy" (as far as these things go)—all we need to do is measure the redshift[25] of light from the galaxy in question. On the other hand, determining the distance to objects out in space is nontrivial and requires a whole toolbox of tricks that build off each other as we measure distances farther and farther away. We call this the "cosmic distance ladder," with each "rung" of distance bootstrapping from the previous. We end up with lots of plots that show that galaxies farther away from us are moving away from us faster.

Although we knew the universe was expanding, we also thought that surely this expansion must be slowing down. Given that the universe is full of stuff that has mass-energy, and that mass-energy brings its friend gravity, and gravity has an attractive force, things ought to be slowly decelerating. This is similar to throwing a ball up into the sky—as the ball is going up it will continuously slow down until it eventually starts to fall back to Earth. In the case of the universe, the question at hand was whether the expansion was decelerating fast enough to cause the universe to eventually recollapse, or whether the universe might just keep expanding forever—like throwing that ball fast enough to escape Earth's gravitational field.

Two competing teams set out to determine how fast the expansion of the universe was decelerating. To do this, they needed to get accurate distances and velocities of galaxies as far away in the universe as possible, which in turn required enormous telescopes. If the expansion is slowing down, that means that as we look farther away galaxies should be moving away from us even faster than galaxies that are closer to us in the universe—because the universe was expanding faster in the past (and as a reminder, the farther away we look, the farther back we are observing in time). As a result, we would predict a deviation from the slope of the Hubble constant in the relatively nearby universe.

Being an astute reader, I'm sure you can guess that the story doesn't end as we expected. What the experiments found was utterly paradigm-shifting; not only was the expansion of the universe not slowing down, *it was speeding up.*[26] This is like throwing that ball up into the sky and having it *accelerate away from you.* What the two teams found looked like this:

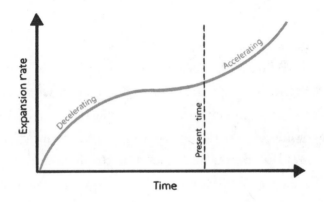

If only one group had gotten these results, the broader community would likely have been even more skeptical than we were. The fact that two independent teams, using different telescopes and their own internal methods, got the same results signaled something else was going on, but what? My sense is that the overwhelming feeling among astronomers at the time was that something was wrong with one of the steps in the distance ladder, and whatever assumptions we were making didn't hold in the distant universe. A flurry of activity ensued over the coming decade, and some minor adjustments were made, but as study after study came in adding more credence to the established methodologies, we were forced to conclude that the expansion of the universe was, in fact, accelerating. But why?

An utterly remarkable feature of the observations that we have now amassed is that we appear to be living at a time when the expansion of the universe is transitioning from decelerating to accelerating, or in other words from being dominated by matter to being dominated by dark energy. Ever since Copernicus kicked the Earth out of the center of the universe, we have learned to eschew notions that our place in space or time could be particularly special, and instead embrace the principle of mediocrity. Nevertheless, we must follow the data.

What is the source of this so-called dark energy? The full extent of our current constraints on dark energy are: (1) it appears to be uniform in every direction, (2) it appears to be getting stronger with time, and (3) it doesn't appear to interact with anything besides gravity. These crumbs of information are not a lot to go on, so the theorists really have a lot of latitude; the main constraint to rein in theorists is the need to be consistent with the rest of physics as we understand it.

As it happens, Einstein had already dealt with a related problem, albeit before we knew the universe was expanding. When he was trying to figure out the physics of space-time, he had a nasty problem with gravity—not dissimilar from what we just encountered. At the time when Einstein was working on this, the observations in hand suggested that we had a "steady state" universe—neither expanding nor contracting, just existing in perpetuity. But because the universe is made of stuff with gravity, the default equations Einstein was working with were unambiguous—existing in steady state was not an option. Famously, Einstein inserted what we call a "fudge factor" into the mix, which in math and physics is code for "We really can't justify this on physical grounds, but this seems to be necessary to get along with reality." Einstein used the Greek letter lambda (Λ) and it appears in the third term of what is known as the "Einstein field equations" as follows:

$$R_{\mu\nu} - \frac{1}{2} R g_{\mu\nu} + \Lambda g_{\mu\nu} = \frac{8\pi G}{c^4} T_{\mu\nu}$$

Since the discovery of dark energy, Λ has had a resurgence, taking center stage in our efforts to understand cosmological observations, and indeed embodies one of the leading hypotheses being considered. This brings us to a list of the explanations being considered (and the problems with them).

Reinstating a cosmological constant is probably the frontrunner in terms of explanations for dark energy. Although we should keep in mind science doesn't proceed via democracy, so it doesn't really matter how popular this idea is in the end if it doesn't match observations. One way to think of Λ is as energy embedded in the fabric of space itself, and the more space there is, the more of this form of dark energy there should be.

To understand how a cosmological constant might work, we need to revisit Einstein's field equations and tackle a pervasive misunderstanding about gravity. This misunderstanding is imprinted on us both through our lived experiences, and through formal schooling—we are generally taught (and experience) that gravity is an attractive force. So before going any further, I want to shine a light on the right-hand side of the equation above, in particular $T_{\mu\nu}$, which is called the "stress energy tensor" ("tensor" is just kind of a fancy math word for "matrix," which itself is a fancy word for a table of values). There is some very cool physics embedded in this term that I don't want you to miss out on, and you don't need to understand how tensors work to get the main point. Moreover, you can casually mention in conversation that you were learning Einstein's field equations. If we expand this tensor into its full form, and we simplify to a "perfect fluid" (which we think is appropriate for space), it is a four-by-four matrix with mostly zeros, except along the diagonal.

$$T_{\mu\nu} = \begin{pmatrix} \rho c^2 & 0 & 0 & 0 \\ 0 & P & 0 & 0 \\ 0 & 0 & P & 0 \\ 0 & 0 & 0 & P \end{pmatrix}$$

Two things I want to point out are (1) that the upper left term (ρc^2) looks suspiciously like mc^2, which might be even more recognizable if I wrote "E=" in front of it (ρ is the variable we generally use for mass density). This ρc^2 term comes from the mass-energy of the universe (i.e., gravitational attraction). And (2) the other diagonal terms are all P which stands for "pressure." Normally, we think of pressure as having a positive value—for example, the pressure inside of a balloon is positive and keeps the balloon inflated (or if you are tired of balloon analogies at this point, you can think of tire pressure, but tires are really just doughnut-shaped balloons). However—and this is the key point—pressure can also be negative. Negative pressure comes from tension, like the actual balloon (or tire) material being stretched and under tension from the air pushing out. Negative pressure acts like negative gravity. If there is enough negative pressure, repulsive gravity can overpower the attractive gravity of the mass-energy term (if $3P + \rho c^2 < 0$). The crux is coming up with a source for the negative pressure. In this (admittedly simplified) scenario, one option might be the that the very stretching of space-time itself could invoke a negative pressure.

There is a glaring problem with this scenario, though—so big that it has been called "the worst theoretical prediction in the history of physics," and once again quantum mechanics is the reason we can't have nice things.[27] Even in its "empty" state, quantum mechanics tells us that the universe has a vacuum "zero-point" energy—in other words, even in the lowest possible energy state, the energy level must fluctuate and can never be zero. The good news is that this vacuum energy should cause the expansion of the universe. On the other hand, these predictions have been wrong by a factor of 10^{120} (although other work has possibly lowered this to "only" 10^{54} orders of magnitude, which I suppose is progress).[28] This problem is known as the "vacuum catastrophe," which alternately strikes me as a solid name for a nerdy music group or a massive recall by the Dyson factory.

There is another possible issue with the cosmological constant solution for dark energy; a cosmological constant might be incompatible with string theory, which boils down to it having a constant value in space-time.[29] Of course, string theory itself remains unconfirmed, but if we find ourselves in need of something hypothetically compatible, we need another option that doesn't have "constant" in the name. Which brings us to another main candidate for dark energy—quintessence.

The quintessence hypothesis gets its name from thinking about dark energy as a fifth fundamental force (hence the "quint"). The primary functional difference between the cosmological constant solution and quintessence is that the cosmological constant Λ is constant—heck, it is right there in its name. In contrast, quintessence is what we call a "scalar field." This is one of those terms that sounds more intimidating than it really is after we translate it to normal human experience. All the term "scalar field" means is that at any given place and time, there is a value of something at that location. For example, if we think about the surface of the Earth, we could assign every location with a value corresponding to its distance above or below sea level at this time—this is a scalar field. Also note that this scalar field can change over time as geological forces do their thing.

As a scalar field, quintessence can change—which makes it compatible with string theory. The ability of quintessence to evolve also might provide an explanation for why we appear to be at a special time when the universe is undergoing a transition from decelerating to accelerating. Specifically, there are quintessence models that have what are called "tracker fields" that manifest differently depending on whether the universe is—for example—matter or radiation dominated. In this case, both our existence and the onset of dark energy might be connected to a phase transition in the universe, making it unavoidable that we would find ourselves living at a "special" time.

In principle, observations may be able to discriminate between different dark energy scenarios, including a cosmological constant and quintessence. These two models have different equations of state that relate core variables, which in turn predict subtle differences in several observables. Among the differences, we can look at the properties of the cosmic microwave background and whether the expansion rate of the universe has been completely uniform in space and time. However, the predicted differences require extraordinarily precise observations to tell the two scenarios apart. We're working on it. At the time of this writing, the Rubin Observatory (named after Vera Rubin) is about to see first light, and there is a major initiative underway to use the new capabilities of Rubin to help us understand dark energy.

Nevertheless, there is a lot we don't understand about dark energy, where by "a lot" I mean pretty much everything about it other than the fact that it exists. So, there is a fair amount of wiggle room for our predictions of the fate of the universe to change as our understanding evolves. For example, if dark energy is due to some type of phase transition (like we think triggered cosmic inflation in the early universe), perhaps the universe will settle down again and this accelerating phase is just a short-lived tantrum. In fact, there are conjectures that this bout of inflation may just be part of a bigger cycle, in which the universe expands to the point that density fluctuations are suppressed and entropy is diluted, after which the universe recontracts to start all over again with a new bang.[30] However, at the current time, it looks like dark energy is not only here to stay, but likely to get exponentially stronger with time unless we can find a good reason for it not to. If dark energy is here for the long haul, this will be a problem for life-as-we-know-it.

The fate of the universe is kind of a big deal. It may seem far off on human timescales (and it is), but just because it won't affect us or our grandchildren doesn't make it less important in the cosmic scheme of things. In general, I wish that we were better at taking the long view. Caring about the world around us today is at least abundant, if not universal. Caring about the universe in the generations immediately before and after we are here also seems fairly easy and pervasive. But it is much harder to feel an acute sense of care and responsibility for distances in time and space beyond a generation or two, which my cynical voice says will ultimately be humanity's downfall. Even if there were some process we could start today, with minor expense and inconvenience, that could provide haven for our distant descendants in a universe being ripped apart by dark energy, would we do it? Given our collective handling of global issues like climate change, I feel pretty secure in my cynicism. Sometimes I wonder whether our universe is, in fact, an experiment being carried out by some superintelligent being to see under what circumstances the inhabitants will step up and prevent catastrophes. If so, I hope there are other incarnations of the experiment that are doing better than we are.

Envisioning what life might become like for our distant descendants—as the acceleration of the universe takes over—is grim. To be sure, I'm highly skeptical that we will survive that long (rewind to the chapter on ET life), but I'd like to give us a fighting chance if we can get our act together, survive our technological adolescence, and learn from the abundance of our historical mistakes. If we can get over these hurdles, our understanding of the universe, technological abilities, and ingenuity might be unrecognizable to us today, or at best only dimly perceptible through the lens of speculative fiction. So, I can't help but imagine what life might be like when (if?) dark energy begins to overwhelm everything else.

As dark energy takes over, the fraction of the universe observable to us will become increasingly small, eventually with no other galaxies within our horizon as they are pulled away. Gravity will be the first force to be overwhelmed and matter will be dispersed throughout space—eventually stopping the formation of stars. In the meantime, our galaxy itself will be pulled apart, as well as the solar system. Eventually, all stars will burn out, and even the black holes will evaporate. Any wavelengths of light still within our horizon at this point will get stretched out to such lengths as to be nonexistent. The universe will be unimaginably cold and dark, with nothing in our horizon other than (perhaps) quantum fields and virtual particles popping in and out of existence.

Unless we are wrong.

Interstition

The previous two chapters have had a tangible quality to them; there are things we can actually observe to shed light on the questions at hand. As such, these last few mysteries are imminently potentially solvable. Over the course of the coming chapters, our path will begin to encounter mysteries that are both more abstract and may also be out of human reach (at least in the near term). In each case, we will start with what we think we know, which serves as an intellectual basecamp from which we can explore deeper into the wild.

Our first sojourn takes us into the realm of black holes. While dark matter and dark energy involve the warping and stretching of the very fabric of space-time, the unintuitive behavior of space-time in black holes becomes acute. Because our understanding (or lack thereof) of black holes is inexorably entwined with the properties of space-time, what happens inside them has connections to virtually every other chapter in this book. We have already encountered how black holes may be related to the Big Bang, and they will also have immediate connections to the nature of time and the possibilities of other dimensions in the subsequent chapters. Curiously, as you will see, black holes might also be connected to why we have the laws of nature that we do and whether the universe is fine-tuned.

Let's see how far we can take the physics that we think we understand, and then ask what might happen in this terra incognita.

6

What Happens Inside Black Holes?

Beware: Black Holes Ahead

Black holes are a perennial favorite topic, and for good reason—they challenge us to contemplate the limits and possibilities in the universe and invite us to consider well-worn plot devices in science fiction as germane to reality. I hope it doesn't come as a surprise that black holes may also challenge you to think about concepts outside of your normal reality.

Curiously, our understanding of black holes shares several confounding characteristics with the Big Bang. Both have a "horizon" that we can't fathom a way to see beyond. Both involve extreme warping or stretching of space-time. And both appear to require a central singularity where the laws of physics, as we understand them, break down.

For most of the objects and places in the universe, we can blissfully go about understanding the physics of how things work without worrying about the dissonance between the two most notoriously nonintuitive concepts in modern physics—quantum mechanics and general relativity. Typically, one of these is so dominant that we can effectively ignore the other. However, the two key situations in the universe for which we can't bury our heads in the sand are the Big Bang and black holes. When talking about what happens inside black holes, we must confront the fact that something needs to give with one (or both) of these pillars of modern physics. Physicists around the globe have been toiling away at a theory of "quantum gravity," but nothing has passed muster yet.

Even before we approach the "center" of black holes (where our understanding of the laws of physics is not on solid ground), reality can get mind-bindingly weird. In my many years teaching about black holes, I have found that different ways of thinking about their weirdness are unpredictably better or worse for different people, so extend yourself some grace when grappling with the concepts in this chapter. We will lean into a series of analogies to help with intuition and insight. However, annoyingly, analogies have their limits, and in some ways they can also be misleading. I'll do my best to flag some of the most potentially misleading bits. I will also unapologetically use *some* math[1] when I think it can help with intuition. If you want more math, I'll be over here cheering you on while you pick up one of the hundreds of books on general relativity available, or if you want to read one of the technical books on anti–de Sitter/conformal field theory (AdS/CFT) correspondence.

This may also be a chapter that you need to slow down and puzzle through things, maybe even loop back. I recall a high school teacher of mine once chastising my class for our shallow reading of a book. My high school self was, of course, offended. But then he went on to say something (ridiculously obvious in hindsight), that blew my mind—it is OK to read things more than once to solidify your understanding. My middle-aged self is like, "Duh." I just want to make sure you know that if you need to read this chapter (or any of the chapters) more than once, you have my full support.

Common Misconceptions

I want to clear up a few things before we proceed. There are a couple common misconceptions about black holes that could get in the way of your understanding, and I want to nip these in the bud. There are two statements that I have found many people believe about black holes that need some gentle correction.

First: *"Black holes suck everything in."* What would happen to us on Earth if the Sun magically turned into a black hole? (For the record, this won't happen; the Sun is not nearly massive enough to do this on its own. I suppose it is possible that another really massive object could merge with the Sun, forcing the Sun's mass over the tipping point, but if that were to happen, we would have serious problems long before we got to the point of having our very own pet black hole.) The short answer is basically, nothing. I mean, of course there would no longer be any sunlight, and unless we found a way to harness immense amounts of energy from another source, the plants would die, and then the animals that eat the plants would die, and so forth. Maybe the giant tube worms we encountered in the chapter on ET life (which get their energy from deep sea vents), would go on to take over the world. But in terms of Earth's orbit around the black hole formerly known as the "Sun," we would be fine. As far as we are concerned out here in orbit, as long as the mass of the former Sun and the mass of the newly minted black hole are the same, the gravity determining Earth's orbit is the same. It is just gravity doing what gravity normally does. To be pedantic, "sucking" suggests a pressure differential between two regions. For example, when you suck liquid up a straw, you do so by creating a lower pressure region in your mouth, and the higher pressure in the outside world forces the liquid up the straw. This is just not how the outside world is affected by the presence of a black hole.

Second: *"Black holes are all really massive."* This misconception is totally understandable; I think there is a tendency to conflate "density" and "mass." Yes, black holes are super dense, by which I mean possibly infinitely dense at their center. But something doesn't have to be "massive" to be "dense," it just must be compressed to a small enough volume. Anything could potentially become a black hole as long as it was compressed down to a small enough size. Given that 99.9999999 percent of the volume of an atom is "empty" space,[2] one can start to conceive of how things that we consider to be "solid" might be compressed down. When I was first learning about black holes, I remember squeezing a rock as hard as I could and trying to comprehend how it could possibly be compressed any more at all, let alone down to the size needed to become a black hole. I decided it would take someone with a grip much stronger than mine.

Now that we've talked about what black holes are *not*, let's talk about what they *are*. The most conspicuous concept tied to black holes—namely that they are "black"[3]—is not in itself so terribly weird. The gist is that the gravity of black holes is so unfathomably strong that not even light can escape. And if light can't escape, well then, of course they appear black.[4] Gravity of that magnitude is clearly outside of our lived experience (for which you might be grateful), but we are generally familiar with the concept of gravity. The only real weirdness so far is getting our minds wrapped around that *much* gravity.

The recipe for generating gravity this intense is straightforward (albeit more challenging to implement), and you can start with any mass you happen to have on hand in the cupboard. Once you've selected your object, all you have to do is compress it down to a sufficiently small size, and *voilà!* OK, fine, that last part may need specialized equipment that doesn't yet exist on Earth, but the *recipe* is easy, in principle.

Generating a black hole comes down to the concept of "escape velocity." Everything with mass has an escape velocity, which for the record includes you, me, and your water bottle. I'm happy to say that the term "escape velocity" pretty much means exactly what it sounds like—the velocity necessary to escape the gravity of something. For example, if you were to throw a ball in the sky, how fast would you need to throw it for the ball to escape Earth's gravitational field? The escape velocity from Earth is about 11 km/s (or 25,000 miles per hour); if you can throw a ball that fast, I hope you have a career in baseball.

This is one of those times when a bit of math can help, so we are going to play with one friendly equation, and I promise no one will get hurt. The escape velocity of an object only depends on its mass and size:

$$v_e = \sqrt{\frac{2\,G\,M}{R}}$$

Hopefully, the variables here are fairly transparent: v_e is the escape velocity, M is the mass of the object from which you are try-ing to escape, and R is the radius from the center of that same object. One of the saving graces of physics is that, in general, people try to use variables that kind of make sense insofar as possible. Usually. Don't worry about G for now—that is just the "gravitational constant," which is just a big number with strange units that we can plug in later. We can use this equation to determine the escape velocity of pretty much anything.[5]

If you were so inclined, you could use this equation to determine the escape velocity of the Moon (~2.4 km/s, quite a bit less than the

Earth's) or the Sun (~620 km/s, quite a bit more than the Earth's). People who care about sending missions around the solar system also care a lot about the relative escape velocities—it is way easier to launch rockets from the Moon than the Earth. With this equation you can start to build up your intuition; because the mass of the object (M) from which you are hoping to escape is on the top, the larger M is, the higher the escape velocity will be. Conversely, because the radius (R) of the object is on the bottom, if the distance from the center of the object is larger, the escape velocity will go down.

If we leave the solar system and start considering more exotic objects, we can dial this escape velocity up to insane levels. For example, "neutron stars" are essentially the densest state of matter possible without being a black hole. Neutron stars are utterly fascinating in their own right, and sometimes I feel that they get neglected just by virtue of "not quite being black holes," but they are as extreme as things can get while still being observable. They are called "neutron stars" because their material has been compressed down so severely that the protons and electrons, which normally kind of do their own things, have been forced to combine into neutrons. These stars are effectively balls of neutrons sitting shoulder to shoulder, which will give New York subways some healthy competition. Speaking of cities, an entire neutron star could fit within the Washington, DC, beltway. And to give you a sense of their density, a single teaspoon of neutron star material would be about 1 billion tons. I don't know about you, but I can't even really conceive of a billion tons, let alone a billion tons in a teaspoon. The point is that neutron stars are really dense, and if we plug the appropriate mass and radius into this equation for escape velocity, we get a whopping ~100,000 km/s, which is about one-third of the speed of light. I don't care how fast your baseball pitch is, that ball is not going to make it off the neutron star. In fact, when it falls and lands on the surface of the neutron star, the ball's impact will have the energy of a thermonuclear explosion. For the record, barring some cool new advanced technology, neutron stars would be terrible places to vacation.

The situation on neutron stars is intense enough compared to the relatively mellow conditions on Earth, but things get extra interesting if we consider what would happen if an even more extreme object had an escape velocity greater than the speed of light. Let's take that equation above for escape velocity and rearrange it to isolate the radius R all by itself:

$$R = \frac{2\,G\,M}{v_e^2}$$

Instead of just any old escape velocity, we can substitute the speed of light (c) for v_e. Now what we have is the radius from the center of the object that is the closest light can get and still escape.

$$R = \frac{2\,G\,M}{c^2} = Schwarzschild\ radius$$

To be fair, this relatively simple equation only works for the simplest type of black hole,[6] and if we add rotation and/or electric charge things get a bit more complex, but the basic principles are the same. Generally, we refer to this boundary beyond which light can't escape from any black hole as an *event horizon*. Astronomers use the word "horizon" quite a bit, by which we mean something we can't see past, or for that matter get any information from whatsoever. "Event" in this case just refers to the reality that we can't know anything about events that happen past this horizon.

To make a vanilla (i.e., Schwarzschild) black hole, all you need to do is compress something down to within its Schwarzschild radius.[7] Generally, when folks refer to the "size" of a black hole, they are referring to its Schwarzschild radius. *Anything* could become a black hole if you could figure out how to compress it down to this size. For example, if there were some way to compress the Earth to within about one inch, it would become a black hole. Look around you at an object that is roughly one inch in size and imagine the entire mass of the Earth in that tiny volume. I suspect that a normal reaction to this thought experiment is along the lines of "But ... how???" I think back to that rock I was trying to squeeze as a kid, and I have a hard enough time conceiving of getting even that rock down to within an inch. This is a prime example of forces in the universe that defy human comprehension. How is this compression possibly accomplished?

If you push down on a table, the table pushes back and resists your push. Have you ever thought about what exactly is doing the pushing back? After all, 99.9999999 percent of that table (and you) are empty space.[8] Given how much empty space there is, one might expect that we could just pass through objects like a ghost. The "pushing back" comes from the resistance of electromagnetic forces holding molecules together.

However, the denser an object is, the stronger its gravity will be. We can force an object to have more intense gravity by either making it smaller or adding more mass. If we keep cranking the values for mass up and/or radius down, eventually gravity can overpower the forces pushing back. Once this happens, the object finds itself in a feedback loop: the higher the density, the stronger the gravitational pull. The stronger the gravitational pull, the more the object collapses. The more the object collapses, the higher the density, ad nauseam. If the object has enough mass to crash through a state called a "neutron degeneracy" (which is, perhaps not surprisingly, the state neutron stars are in), there are no known forces to stop the collapse. In other words, the collapse is infinite, resulting in an object that is infinitely small and has infinite density; we call this a "singularity." At least this is what happens according to our current understanding of physics.

Here is the thing about singularities; when we encounter these in math or physics, they are often (usually?) a sign that we have done something wrong, or we don't *really* understand everything at play, or we have oversimplified the physics. The situation with black holes is no different. *As far as we know*, there is no known force to stop the collapse of the object, but there is also a lot we know we don't know (not to mention the things we don't know we don't know). One of these things we know we don't have a handle on is quantum gravity. Most of the astrophysicists I know give it even odds that—somehow—quantum mechanics will come into play and help us avoid an embarrassing singularity in our math. There has been a flurry of activity trying to solve this problem, and there are a couple theories in play that might help us out—"string theory" and "loop quantum gravity," which might result in things like holograms on the surface of black holes, or black holes actually being "fuzzballs" (I'm not kidding, that is what they are called), but I want to put a pin in those ideas and come back to them later.

The black holes that result from this process of gravitational collapse can come in a range of sizes. Perhaps the most common mechanism for black hole formation is through the death of the most massive stars, creating black holes with masses of, say, ten times the mass of the Sun (or so). We also have very strong evidence that supermassive black holes are lurking at the centers of virtually all massive galaxies, and these monsters can have masses in excess of a billion times the mass of the Sun. *Exactly how* these behemoths form is an active topic in modern research. All the black holes that fall between these two categories of "stellar mass" black holes and "supermassive" black holes are shockingly called "intermediate mass" black holes. There are also predictions for *primordial black holes* that might have been created in the intense conditions of the very early universe; these cuties would be smaller than the width of a fingernail with masses similar to that of Earth's moon. It is even possible that if a couple subatomic particles smashed together hard enough their mass-energy could be compressed to within the relevant Schwarzschild radius and create a particle-mass black hole, which could, in principle, even happen in a particle collider on earth.[9]

Of course, regardless of how a black hole emerges, weird things start to happen in their vicinity. Before we even get inside a black hole, we are forced to confront conditions that call our everyday sense of reality into question. The first effect you might notice is what is called "gravitational redshift." I feel fairly confident in assuming that you have, at some point, traversed up a hill or staircase. It took energy for you to do this. Heck, we literally put StairMasters in gyms for folks to use energy in this way—without even actually making upward progress (sometimes I think our distant ancestors would find the things we do utterly insane, like building moving staircases that don't take you anywhere). Of course, the reason it took energy to climb the hill or staircase is that you were actively fighting gravity. You can imagine that if you just kept climbing, you would use more and more energy, until perhaps finally you just ran out of steam and sat down.

The same thing happens to light. For light to move away from something with gravity, it must use energy. Since light doesn't have the luxury of burning stored-up fuel like us humans, the only place for that energy to come from is how fast the light wiggles—or if you want to be slightly more scientific—the frequency of the light. If light wiggles less, it has a lower frequency. As light climbs out of a gravitational well, it loses energy, and its wavelength gets increasingly longer, which means redder. For the record, this happens to light climbing out of the gravitational potential of anything, the effect just isn't as pronounced when that "anything" is like an apple. As a result, any light we might be lucky enough to receive from near a black hole will be "redshifted." The deeper in the gravitational well the light starts from, the more energy it takes to climb out. Fine, that makes sense, we've all gone up hills. Here is the kicker: if light tries to leave from the event horizon of a black hole, it would take an infinite amount of energy for it to get away. If that light loses an infinite amount of energy, its wavelength would be stretched to infinity—which means it doesn't have a wavelength at all, at which point one could argue that light no longer exists.

As a side note, the inverse is also true—as light falls into a gravitational potential well it *gains* energy and is "blueshifted." So, if you happened to be hovering near the event horizon of a black hole, you would be bombarded with ultrahigh energy radiation from the surrounding universe. Normal sunscreen is not going to cut it.

The next bit of weirdness you might experience near a black hole is "spaghettification."[10] Here's the deal—the force of gravity is stronger the closer you are to the source of the gravity, and black holes are no exception. The thing with gravity is that it obeys what we call an "inverse square law," which means that it gets stronger or weaker with the *square* of the distance from an object with mass. Because of this square law, the gravitational force on the side of an object closer to the black hole is way stronger than the force on the opposite side. Or to make this personal, if you were to jump into a black hole feet first, the gravitational force on your feet would be significantly stronger than on your head—the result of which is that you would get stretched out. As you approached the central black hole and the gravitational force gets increasingly intense, this would eventually pull apart your molecules and stretch you into a long string . . . kind of like spaghetti.

So, could you survive falling into a black hole? I'm so glad you asked. Here is (yet another) fascinating thing about black holes that is perhaps (yet again) counterintuitive: the more massive a black hole is, the gentler crossing the event horizon would be. The radius of an event horizon for a Schwarzschild black hole gets larger as we add mass to the black hole, which is not surprising—the more massive a black hole is, the bigger its event horizon. But the radius of the event horizon increases *linearly* with the mass—just take a quick look back at the equation for the Schwarzschild radius to convince yourself of this. If you double the mass, you double the radius. This starts getting counterintuitive when we think about *density*. Density changes proportionally to the *radius cubed*. When we put these two behaviors together, it means that as we add mass to a black hole, the average *density* goes *down* by a factor of the radius squared. By the time a black hole becomes "supermassive," the average density within the Schwarzschild radius could be similar to the density of water, which really doesn't seem so intimidating to me. This means that you could hypothetically cross over the event horizon of one of these gentle giants without being stretched into a string of molecular spaghetti.

In addition to spaghettification, there is another—very hypothetical—problem with entering a black hole. I have my own doubts about whether this conjecture might be correct, but I am impelled to include it here just in case it is, and you are planning a trip to a black hole. One solution to the "information paradox" involves a "firewall" just outside the event horizon of a black hole that would incinerate anything falling in.[11] (The "information paradox" is coming up in a few pages; feel free to jump forward and have a look. Do you remember "Choose Your Own Adventure" books from when you were a kid? It is tempting to invoke that device in this book, which necessarily references topics forward and backward.) So even if you didn't get spaghettified, you might want to rethink your planned visit.

Which brings us to the third "fun" thing about approaching a black hole—it would seriously mess with your experience of time. We've reached the point in talking about black holes that departs from anything that is likely to make intuitive sense. THAT'S OK. Think about where our intuition comes from—the integral of our past experiences woven into an understanding of reality. I happen to be a big fan of intuition—I often feel intuition gives me insight into things I haven't been able to consciously articulate to myself. But I also know that intuition has its limits and—in particular—trying to apply (or expect) intuition in conditions that we humans have never experienced can lead us radically astray. My ardent request to you, dear reader, is not to shut off your brain just because something doesn't make sense. This is a plea I make to my students as well. Try to hang on to whatever bits and pieces you can, and eventually—hopefully—you can at least make sense of why things don't make sense.

I am asking you to cast aside the essential notion of time in your lived experience and replace it with the following:

The notion that time passes the same for everyone is wrong.

In the corner of the universe where humans have evolved, we simply do not encounter scenarios in which we would experience time otherwise. But just because we haven't experienced it, doesn't make it not so. After all, the environment on and around Earth is like the milquetoast of the universe. What modern physics tells us, and experiments verify, is that the rate at which time passes for you depends on your speed and the gravitational field you are in. You can blame Einstein and his theories of special and general relativity.

Here is an imaginary scenario of how this might play out:

Suppose you have a friend with no sense of self-preservation and access to a really fancy rocket. In this thought experiment, we've found a black hole in the nearby universe, and your friend wants to go check it out. After reading this book, you know that checking out the black hole might not be the best idea. Sadly, your friend doesn't believe you and thinks scientists just make things up to feel important (never mind that your friend is using a rocket built by virtue of our scientific understanding, but people use the products of science while denying science all the time). You have a terrible fight, and your friend storms off to their rocket.

Shortly after they launch, you start to feel bad and want to make up, so you send them a voicemail, letting them know you care about them and would they please turn around and come home. In the meantime, you can monitor their vital signs from the rocket's downlink, which your friend gave you access to before the big fight. Your friend gets your message just fine but is still mad and wants to prove you wrong, so they don't even bother replying. Their vital signs look normal-ish, maybe the pulse and blood pressure are a little high, as one might expect from the stress of being on a rocket.

Time goes by, still no word from your friend. After a while you notice that their heart rate has gone down, and at first you are relieved—maybe the stress has abated, and they are doing OK. Then you recall reading this chapter of this book again. Uh-oh. Taking a second look, their heart rate is way too low. You send them another voicemail, telling them you love them and begging them to come home. This time, when they receive the message, your voice is super shrill and you are speaking so fast they can barely understand you. Your friend sends a message back, telling you everything is great and they are fine, but why are you speaking so quickly and in such a high pitch?

When you get your friend's voicemail, their speech is stretched out and at an ultralow pitch. To even understand what they are saying, you have to play the audio at high speed. You know exactly what is going on and check your friend's vital signs to confirm: by this time their heart is only beating once per minute. You send your friend one last voicemail as a Hail Mary, but this time they receive just an infinitesimally brief and excruciatingly high-pitched tone. They try to send you a message asking why you are torturing them with high-pitched noises. They are fine, and you should just chill out. But why are the planets going around the Sun so fast? You never get the message.

You watch their vital signs over the coming weeks, then months, then years. Their heartbeat just goes down and down and down until it doesn't even beat once in a whole year, and with high-powered telescopes you can see that their rocket now appears frozen in space. Your friend, on the other hand, just watched the Sun expand and die, followed shortly by the collision of the Milky Way with the Andromeda Galaxy, and eventually the end of the universe.

The moral of this story could be to not visit black holes. Or it could be to try to help promote scientific literacy. Regardless, the point is that time did not pass at the same rate for you and your friend; your friend experienced what we call "gravitational time dilation." In a nutshell, the more intense the gravitational field you are in, the slower time passes for you relative to an observer sitting comfortably in a low gravitational field. Or, equally valid, if you are in an intense gravitational field, you observe time passing more quickly for people or processes happening in a low gravitational field. To be sure, this isn't just physicists making things up that are fun to think about. This is experimentally confirmed—for example, the GPS satellites, which operate in a lower gravitational field than what is found on the surface of the Earth, have clocks that tick 38 microseconds faster per day than clocks on Earth. I know that 38 microseconds doesn't sound like a lot of time, especially when you consider how many microseconds you spend doing other things (I estimate that you will spend roughly 30 billion microseconds reading this book). But in the realm of relativity, what matters is how fast light can travel in 38 microseconds, which is more than a kilometer. Even a minuscule offset from gravitational time dilation can manifest in macroscopic effects on Earth.

This brings us to the "warping" of space-time due to gravity, and I happen to be the kind of nerd who thinks this is genuinely fun to think about. One issue is that visualizing more than three dimensions of space and one dimension of time at once is hard, maybe impossible. There are solid evolutionary reasons for this—we simply haven't needed to perceive or visualize more dimensions for our survival. This doesn't mean that there are not more dimensions out there, just that we don't interact with them (which is the topic of the chapter on dimensions). For us humans to build intuition about multidimensional concepts, we often need to come up with tricks and analogies. One of these analogies that has helped scores of people better understand general relativity is the concept of "embedding diagrams."

We normally think of space as having three dimensions (up-down, side-to-side, and front-to-back). The trick with the embedding diagrams is that we ignore one (or more) dimension of space—for example, we can pretend that up-down doesn't exist, and we live on a flat sheet. In this 2D world, we can conceive of side-to-side and front-to-back, but up-down has no meaning. If you asked someone "What's up?," it would be nonsensical, and you would have to say something like "What's on the side?"

Once we are comfortable thinking about space (or space-time) as a two-dimensional sheet, we can invoke the real power of embedding diagrams, which uses that discarded third dimension to help us visualize warping. In astrophysics classrooms around the world, students are asked to visualize this 2D fabric of space-time as a rubber sheet stretched across supports.

When we put something massive on the rubber sheet it will warp; the warping will depend on the shape, size, and mass of the object we put on it. Moreover, and this is an essential bit for what follows: light travels along the rubber sheet—so if the rubber sheet is "warped," so is the path of the light; light will always go in a straight line in curved space-time (straight lines in curved space are called "geodesics"). Notably, a straight line in curved space-time looks like a curved line in flat space-time.

In many ways, this is a great analogy for gravity—objects with mass warp this rubber sheet of space-time. This analogy has helped vast numbers of perplexed students build intuition. But this analogy has its limits, and these limits can cause confusion if they aren't pointed out.

One limit of this analogy is this "extra" dimension that the rubber sheet is warping into. We don't necessarily think that there is another dimension into which space-time is warped, but rather that the coordinates we use to describe what happens to spatial and temporal durations behave in a way that we can visualize with warping.

A related breakdown of the analogy is that the warping of the rubber sheet implies that there is some source of gravity outside of the rubber sheet pulling the massive object down. Instead, the warping from objects with mass has to do with how objects warp the coordinate grid of space-time around them, which actually happens in full three-dimensional glory (i.e., not downward, but rather inward).

With those caveats in mind, we will now invoke embedding diagrams...

The Great Ant Race

Full disclosure, I know almost nothing about ants and how they perceive the world, although I do find their communal behavior and inclination toward self-sacrifice to be fascinating. If you happen to be an entomologist who specializes in ants (which I've now learned is called a "myrmecologist"), I hope you will excuse my blatant disregard for ant physiology in what follows.

Suppose there are two ants at a picnic, let's call them "Antony" and "Antonella," and this special picnic happens to be on a rubber sheet with a *very* dense and heavy bit of fruitcake in the middle. Being ants, Antony and Antonella can only perceive the two dimensions of the rubber sheet and have no concept of "up" and "down" (I have no idea whether this might actually be true of ants, but let's just pretend for now, OK?); because they have no concept of "up" or "down," they

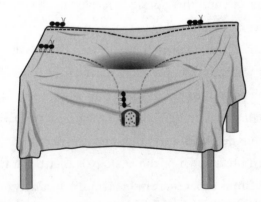

don't perceive the warping of the rubber sheet due to the super-dense fruitcake that, shockingly, apparently no one wants to eat. Moreover (and this is an important bit), in this analogy we must remember that *light travels along the rubber sheet*; this means the two ants see the fruitcake sitting there—it appears to them as if it were just sitting on a flat rubber sheet. In other words, after the light leaves the fruitcake, it travels along the curved rubber sheet until the light hits the ants' large compound eyes.

Enjoying their lovely Sunday picnic, Antony and Antonella decide to have a pre-lunch stroll to work up a good appetite, and they agree to walk from one side of the rubber sheet to the other at the same steady pace. To scout more territory, Antony starts at the midline of the sheet, and Antonella starts on the side. Things start as expected, and the two ants exchange small talk as they stroll. However, soon Antonella thinks that Antony has slowed down. Meanwhile Antony thinks that Antonella has sped up. Bickering ensues. In "reality," each of them is strolling at the same speed with respect to the rubber sheet—they have each walked the same distance *along the rubber sheet*.

The issue is that they don't perceive the warping of the rubber sheet. Antony thinks Antonella is getting faster and faster, while Antonella thinks Antony is getting slower and slower. If that fruitcake is *super* dense and the warping of the rubber sheet extreme enough, from Antonella's perspective, Antony will appear to just stop making any forward progress at all. From our outside omniscient perspective, we can see that Antony stopped making steady progress toward the right because he is now headed "downward" with his even gait, but "downward" means nothing to these venturesome ants.

This is essentially how gravitational time dilation works, and the appearance of Antony not making forward progress is analogous to how time dilation behaves at the event horizon around a black hole.

The key point to remember is that there can be a significant difference between what we perceive and reality.

Another tool we invoke when thinking about relativity (both special and general) to build intuition about space and time near black holes are visualizations conveniently called "space-time diagrams." With these diagrams, we limit ourselves to a single dimension of space in addition to the dimension of time, which helps us see how these two dimensions play off each other. I'm going to take some liberty here with these diagrams in contrast to how they are typically used in physics to illustrate a couple points that might help with intuition.[12]

You have probably been told at some point that light travels at the speed of light (and the name kind of gives it away). And also, that you are bridled to the plodding speeds of things with mass, which can't go the speed of light. This is true, as long as we are talking about the speed of light exclusively *in space*. However, when we invoke the framework of *space-time* there is an extraordinarily striking feature

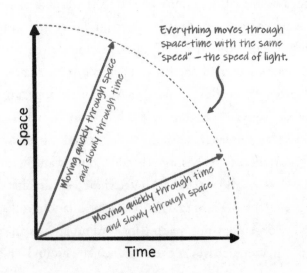

Everything moves through space-time with the same "speed" — the speed of light.

Moving quickly through space and slowly through time

Moving quickly through time and slowly through space

Space

Time

to point out: you, light, and a rocket are all moving at the same speed as light through *space-time*.[13] The length of vectors in this version of space-time diagrams are all exactly the same.

The crux of this is the following: the faster you move in space, the slower you move in time, and vice versa. This concept is at the core of special relativity. To be clear, if you are sitting still *in space*, that means you are moving at the speed of light *through time*. And the flip side—anything moving at the speed of light in space (e.g., light), it is sitting still in time. In other words, light does not "experience" time (I recognize the ridiculousness of anthropomorphizing photons, but so be it). The upshot is that at this very moment you *are* moving at the speed of light, just mostly through time, which is kind of a bummer if you don't want to get any older.

Similarly, we can play around with versions of space-time diagrams that invoke principles of general relativity. There are some nifty features of general relativity that will come out of this analogy if you can stick with me here.

Let's visit Antony and Antonella and see what shenanigans they are up to. Now at home in their apartment building, Antonella lives on the top floor while Antony has a garden-level apartment. The key to how I've oriented the diagram that follows is that gravity is stronger at the bottom (closer to the Earth), and therefore Antony is deeper in the gravitational well. This means he is moving more slowly *through time*.

Recall that what Antony and Antonella perceive is not necessarily the same thing as "reality." If we map them onto paths in a space-time diagram, it might look something like the following—Antonella is moving faster through time. But to keep the length of their vectors the same in space-time, we need to stretch out the time axis—this is where this thought experiment can get a little challenging, but I promise it is worth it to puzzle through. A key element is that the axis for space and the axis for time must stay orthogonal. Later we will encounter the possibility of space and time not always being orthogonal, but that discussion can wait. Thus, the grid lines of space and time must always meet at right angles.

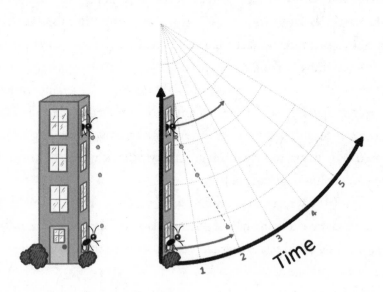

Note that both of these plucky ants are standing on their apartment floors, which are holding them up at the same distance from the ground, so their paths in space-time in this diagram have to curve along with the grid lines of space. Now that Antonella and Antony have vectors of the same length in this diagram, you can read off how much time has passed for each of them along the bottom axis while you've been reading. While Antonella has experienced four ticks of time, Antony has only experienced two.

Both Antonella and Antony travel through space-time at the same rate, meaning their vectors in this framework have the same length. Because Antony is deeper in the gravitational potential well, time is more stretched out for him.

I want to pause and point out something nifty about gravity here. Do you remember learning Newton's first law—that *a body in motion will move at a constant speed in a straight line unless acted on by an outside force*? When we apply this law to our ants' antics, we learn something very interesting.

Antonella wants to share a crumb of food with Antony, but living all the way at the top, she doesn't want to bother to come downstairs. Antonella did well in high school physics, and she knows that gravity will be "an outside force" acting on the crumb, and therefore the crumb will fall down. She decides to toss the crumb out the window, and not unsurprisingly it falls down. Comparing Antonella's perception of what is happening with our newly stretched space-time fabric, something remarkable emerges.

In curved space-time we have flipped the script. In Antonella's perception, the crumb follows a curved path in space-time and appears to accelerate because it is acted on by an outside force—gravity. In our stretched space-time, the crumb *follows a straight line* in curved space-time. In fact, in curved space-time, everything follows a straight line unless acted on by an outside force. By contrast, Antonella herself is now following a curved line in space-time because she is being acted on by an outside force—the apartment floor holding her up! The bottom line is that gravity is just a manifestation of any given object moving in straight lines in curved space-time. I dearly hope that, at least from time to time, you will think about gravity this way—at least insofar as you think about gravity at all, which I also hope you will do more of now.

I want to take this analogy one step further, which may be one step too far for some people, but I can't resist because, frankly, I think it is really cool. We now have this space-time diagram in which we have stretched out the time axis to mimic the effect of gravity on time dilation. Of course, the more intense the gravity, the more we must stretch out the time axis. What happens in this diagram when we get to the event horizon of a black hole? In effect, we take that time axis and stretch it *all the way around*; we've gone from a nice square-shaped diagram to a full disk. When we do this, the top of the diagram gets compressed to a point, and thus even if Antony moves a minuscule amount in time, Antonella would go around and around the point at the top through the whole future of the universe. This is another way you can envision time dilation at the event horizon of a black hole.

But wait, there's more! What if we stretch out time even further? To do this in our nice flat two-dimensional analogy, the space-time diagram has to pop into a third dimension, which is neither space nor time. For example, it could take on a spiral shape into this third dimension. This comes to my mind whenever I try to make a pie crust, which I am really very terrible at; if you roll out the dough too much in one spot the crust can't lie flat on the surface, and it ends up buckling up or folding over. What is this third dimension that space-time has popped into? This is the equivalent of time becoming mathematically imaginary beyond the event horizon of a black hole, and in this sense imaginary numbers are orthogonal to the "real" numbers we are used to dealing with in normal space and time.

At this point, I fully support your going off to make a pie crust and filling it with something delicious. As a bonus, enjoy with friends or family while explaining the concept of space-time.

Now back to black holes.

The Information Paradox

Before we can talk about what happens "inside" black holes, we need to address an elephant in the room (at least in a physics classroom).

You may not have put the concepts of "black holes" and "information" together in a sentence before, but it turns out that the two are entangled in a way that leads to an apparent paradox. In brief, quantum mechanics tells us that information must be preserved, but black holes appear to swallow information whole, removing it from the universe. Either a fundamental aspect of quantum mechanics is wrong (in which case we have bigger issues), or something strange happens with black holes (which I know is shocking).

Like almost all paradoxes, the information "paradox" may not be an actual paradox, but it does reveal an interesting feature of black holes that we need to sort out—this topic brings several nifty threads of physics together into one convoluted knot. The process of trying to untangle this knot is revealing entirely new ways of thinking about the universe. I've added this section here—before we talk about what happens beyond the event horizon—because some of the possible answers about what might happen past the event horizon come from considering the conundrum we find ourselves in. This is also one of the things that made Stephen Hawking famous.

This single topic weaves together an array of physics ideas, which I will add one thread at a time. At first, they may seem disconnected, but they will come together. Also, maybe go get a cup of coffee now.

Here is a Cliffs Notes version to help you orient yourself in what follows (or for younger readers, TLDR): When stuff falls into a black hole, we appear to violate the quantum mechanical principle of the conservation of information, which would overthrow a hallmark of modern physics. We might be tempted to think this information is just inside the black hole and not actually *lost*, except it turns out that black holes can evaporate. If a black hole evaporates, what happens to the information that went in? Resolving this apparent paradox has led to some absolutely bizarre hypotheses.

Now for a more detailed account of this so-called "paradox."

The first step is a concept called "entropy." I am about to try to convince you that entropy is essential. In fact, entropy is so entwined in the universe that it will poke its head up several times in this book to say hello. Because I don't want you to run away (or close the book) screaming when you read the word "entropy," I want to offer a little introduction now. You can either just follow along or, better yet, grab your own deck of cards. Just for fun.

Let's say that you and I both have a deck of standard playing cards, and we want our cards to be in exactly the same order. If I had a brand-new deck of cards, it would look like this:

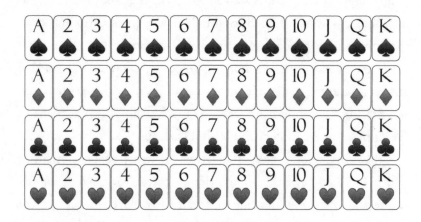

I *could* tell you every single one of the 52 cards in order—indicating both the suit and number for each card (roughly 150 pieces of information total), but I don't need to—the number of statements necessary to perfectly describe the deck to you is quite small. I could simplify our lives by just describing the order of my cards as follows: (1) aces low, (2) spades, (3) diamonds, (4) clubs, (5) hearts.

I feel confident that with these minimal instructions, we could end up with identically ordered decks (as long as we share an understanding of the ascending order of the cards, but if not, I could give that list to you as well). That is kind of boring, though, so let's spice things up—I'll shuffle the deck once (keeping in mind that I am not a very advanced shuffler of cards). Just to prove my level of nerdiness to you, while writing this I got out a deck of cards, put it in the above order, and performed this experiment. You are welcome to play along at home and try for yourself.

Here is the resulting order:

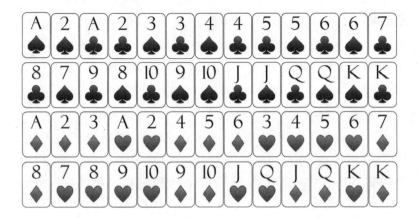

As with the brand-new deck, I *could* list every single card in order (making 52 statements, each with the number and suit), but there are some embedded patterns that I can use to simplify the instructions to get your deck in the same order. I challenge you to see how short you can make a list of instructions to describe this card order exactly. It will be more than the five statements above, but also significantly less than 104 if we list every card's properties in order.

Just to make my point, and because I am sure reading a book about shuffling cards is riveting, I am going to keep shuffling a few more times. After a second shuffle the suit order has started to get mixed up, although notably the order within each suit still goes from low to high, but there are still patterns embedded in this card order. Describing this deck *exactly* would take even more information statements than the version above, but still fewer than 104.

I could keep shuffling for a long time, and with each shuffle I expect more people would close this book and stop reading, but indulge me just once more. On the third shuffle, both the suit order and number order start to get mixed, requiring many more statements to describe the deck order exactly.

We could, of course, keep going, and statistics tells us we should—to get a properly shuffled deck of cards that is indistinguishable from a random order takes seven shuffles.[14] Casinos will use this information to—wait for it—help you part with your money.

What does this have to do with *entropy*? Everything.

Often the concept of entropy is introduced as *the amount of disorder in a system*. I feel like my house is a case study in the entropy lecture, and if my students have trouble grasping the concept of entropy, I joke that they should visit my house on a field trip. With three kids, my spouse and I with full-on careers, and the number of animals typically in the range of four to six, our house often feels like someone turned it upside down, shook it, and then flipped it back over. When we talk about the second law of thermodynamics (abbreviated: entropy increases), I am partially convinced that children are at the root. I mean, why else is there a toothbrush in the shoe closet? That being said, "disorder" feels like such a negative word. How do you even measure the "amount of disorder"?

Another way to think about entropy, which I find to be much more intuitive, is *the minimum number of statements necessary to perfectly describe something*. In the case of the deck of cards, the original deck order had very low entropy—only a couple statements were necessary to completely describe the order of the cards. As we continue to shuffle the deck, the number of statements we needed to describe it increased—in other words, we increased the entropy of the deck.

Full disclosure, I hated studying thermodynamics as a student. I just could not find a way to make the concepts, variables, and plots as intuitive as they seemed to my classmates. I could not keep "entropy" and "enthalpy" straight, and all the phase-space diagrams with pressure and temperature seemed like abstract art with no logic that I could discern. If you happen to want to run away screaming when you hear words like "entropy," I hear you. Maybe if the first time one encountered concepts like entropy they had a positive feeling toward the concept, they wouldn't develop the same aversion I did.

For many years after my first exposure to entropy as a student, my brain would just go into lockdown when I heard the term, and as a result my relationship with thermodynamics has been fraught. This aversion came to a head one semester when my thermodynamics class had a take-home exam with forty-eight hours to complete it. I started right away because I *knew* the exam was going to kick my butt. I always believed that I was the weakest student in my physics class (and you would still be hard-pressed to convince me otherwise), and when it came to thermodynamics there was no doubt in my mind. As I expected, I got stuck on the very first of five problems in the exam (that we had forty-eight hours to solve five problems tells you something about the level of the problems). My heart sank and I lost any remaining sense of fortitude I might have been able to muster. I spent a full twenty-four hours fighting with that first problem and trying to figure out what I was missing.

Finally, with half of the forty-eight hours gone, I moved on to the other problems—knowing that my score was already down by 20 percent after skipping the first problem. I spent the remaining twenty-four hours on the last four problems, with little (if any) sleep and a level of caffeine consumption that would probably send my middle-aged self to the hospital. I fought through, and at least got some semblance of answers I could turn in.

With thirty minutes left to walk it over to the physics building, I decided to take a quick look back at that first problem and see if there was anything I could do to salvage some partial credit. Good grief, I had *misread* the problem. Reading it again, I could see that what I had been trying to solve was impossible. What I was being asked to solve was straightforward. But I was out of time, and that was that. In hindsight, my own expectations going into the exam seemed to be self-fulfilling. I think there is a lesson in there.

I trudged across campus to the physics building, sealed my fate, and hoped the grader had some mercy. I remember staring at the sidewalk as a I walked, with the nice concrete squares, many of them with cracks spidering across in various directions. In my sleep-deprived state, I started to have waking dreams (or hallucinations?)—the concrete squares themselves were the axes of phase-space diagrams, and the cracked lines were mapping parameters in thermodynamic plots. To this day, I cannot look at cracks in squares of sidewalk concrete without thinking about thermodynamic plots.

It took me years after my first exposure to entropy to not feel a fight-or-flight response (usually flight) when thermodynamics came up in my studies (and it has a way of *always* coming up). The good news is that I've mended my relationship with entropy, and it is really important. Now whenever I shuffle cards I feel (a little) warm and fuzzy about entropy.

The next step is to send entropy and a black hole on a date and see if they are compatible. (Spoiler: they are not.) To the outside world, black holes are highly ordered systems, and can be described with just a few variables. Unfortunately, that is not true of intrepid astronauts or random decks of cards. So if they enter a black hole, where does this information go?

The prevailing conventional wisdom has been that any black hole can be described with three pieces of information: its mass, charge, and angular momentum. As the late physicist John Archibald Wheeler famously said, "Black holes have no hair," which became known as the "no-hair theorem." Physicists must have had such a grand time naming things in the last century. I would argue that in this scenario black holes have three strands of hair, but I will concede that a "three-strands-of-hair theorem" doesn't roll off of one's tongue as well. So, what does the hair of black holes have to do with entropy?

Imagine you took this now fully shuffled deck of cards and tossed it into a black hole. While we needed roughly 150 bits of information to describe the deck of cards (the suit and number of each card along with their order), the black hole the cards went into is described by *only* three statements. This means the entropy of the universe outside the black hole *went down* when you offered your deck of cards as a sacrifice. The second law of thermodynamics is now raising its hand to remind us that the entropy of an isolated system (like the universe?) is only supposed to be able to increase with time. Oops.

The first solution that might come to mind is the following: just because we don't have access to the entropy that is now inside the black hole, that doesn't mean that the entropy isn't there lurking out of sight, which totally seems in character with the entropy that lives in my house. That is a fair point, and to be sure this possible solution is one of the first things physicists considered. And it looked like we might have a nice and tidy explanation, at least until Stephen Hawking found a way to preserve the second law of thermodynamics . . . through black hole evaporation. I know, the whole deal with black holes is that nothing can escape, and now we're saying they can *evaporate*? Come on, though—did you really expect black holes to be simple intuitive objects?

The key to black hole evaporation is antimatter. Like entropy, antimatter will come by for a visit a few times in this book, so there are a few key pieces of background to put in your pocket. First, antimatter exists. Second, every type of subatomic particle has an "antiparticle," which is identically opposite (just accept the oxymoron and read on); a given particle and antiparticle pair have the same mass, but opposite charge (like an electron and its antiparticle, the positron). Third, some particles are their own antiparticles—in particular, photons (which are electrically neutral) are their own antiparticle, which strikes me as deeply schizophrenic. Fourth (and this is my favorite part), if a particle and its antiparticle encounter each other, they mutually annihilate and turn into energy.

You may have noticed that our universe seems to be overwhelmingly dominated by normal particles without a corresponding number of antiparticles running around. This is a bizarre asymmetry that I am saving for the chapter on the nature of time. But you can thank this asymmetry between matter and antimatter for your existence; if not for this extreme asymmetry, matter and antimatter would all just mutually annihilate.

The next concept to weave into this framework would admittedly seem outlandish if not for empirical enquiry (aka the scientific method). To be sure, if one isn't familiar with the scientific method, peer review, and how to read and interpret results from competing teams, I can understand how the following concept may not seem all that different from "miracles" like Moses parting the Red Sea, which—for the record—we have not yet recreated in the lab.

Here's the deal: because of underlying quantum fluctuations, particle-antiparticle pairs can pop into existence out of "nothing."[15] I will be the first to agree that this sounds like it ought to violate both common sense and the physical conservation principles that students are forced to memorize. It turns out that nature has a loophole—as long as these particle pairs annihilate within a quantum timescale (i.e., really quickly and before they can interact with anything), they can pay back the energy loan they took from the universe, and our conservation principles just kind of look the other way and go about their business.

We only really have trouble with the conservation principles if the particle-antiparticle pairs don't immediately annihilate. Perhaps now you are wondering what might keep them from annihilating... if only the topic of this chapter offered a clue.

Now we have the background to talk about black holes evaporating (at least we think they do; although we have not yet observed this, theorists are pretty convinced). Let's zoom into particle-antiparticle pairs popping into existence in the vicinity of a black hole. Sometimes both partners will fall in, and as far as we're concerned at that point they are out of sight and out of mind. Maybe they recombine and annihilate, maybe not. We'll never know. Sometimes the partners will annihilate outside the event horizon, and the conservation principles can keep their gaze averted. But *sometimes*, one partner might fall into the black hole, while the other manages to escape. To the outside world, it would now appear that the vicinity around the black hole is emitting radiation (known as "Hawking radiation"). To be clear, the radiation isn't escaping from *within* the black hole. Nevertheless, this radiation has an effect on the black hole, because now we really do have a problem with conservation principles.

The now-separated particle pair took out an energy loan worth two particles from the universe but did not pay it back by annihilating. At this point, the conservation principles can no longer pretend to be ignorant of these particle shenanigans—the loan must be repaid, so the mass-energy is taken from the black hole. The black hole gained one particle worth of mass-energy in the process, but it must pay back two particles worth of mass-energy, and consequently *it shrinks* by one particle worth of mass. *Voilà*, we have a mechanism for black holes to evaporate.

As we have seen, in the vicinity of a black hole, the geometry of space-time is warped, and the more warped the space-time is, the more likely it is for the particle-antiparticle pair to get separated. The lower a black hole's mass, the more extreme the warping near the event horizon, therefore, the more particle-antiparticle pairs will get separated, and the faster the evaporation process happens. To give you a sense of this, if you could isolate the supermassive black hole at the center of our galaxy and keep it from absorbing any more mass or energy, it would take roughly 10^{100} years to evaporate. On the other hand, if there were a black hole with a mass of, say, a million kilograms (this sounds like a lot, but it is just a million liters of water, or part of an Olympic swimming pool), it would evaporate in about a minute.

Hooray! We have found a way to save the second law of thermodynamics, which I'm sure gives you a great sense of relief. If black holes can evaporate, they can return all the entropy the black hole ate up back into the universe where it feels at home. However, I'm sorry to tell you that sense of relief will be short-lived, because now we have a new problem; we didn't just toss entropy into the black hole—*information* went into the maw along with it. So now we need to talk about exactly what information is in this context, and why it creates a new problem.

Concepts like the conservation of mass, conservation of energy, or conservation of angular momentum are typically encountered somewhere in one's education, and not too intuitively challenging— to me these feel a lot like "common sense" because we humans interact with these conservation laws daily. I would love to go down a rabbit trail right now to point out that every one of these "conservation" principles is related to an underlying symmetry in the universe, which we know thanks to the brilliant Emmy Noether who published this work in 1918. But a deep discussion of symmetries will have to wait for another time.

There is a bonus conservation principle you may not have encountered in school that comes from quantum mechanics: the conservation of (quantum mechanical) information. In quantum mechanics we use something called a "wave function" to describe the state of a system, but we can only ever describe the state of a system in terms of *probabilities*. As far as quantum mechanics is concerned, one doesn't know anything about a system beyond the probabilities—until that system is observed. OK, fine. But one of the things that makes quantum mechanics so whacky is that not only do we *not know* the actual state of the system until we observe it (which makes sense), but the state of the system also doesn't exist until it is observed (i.e., interacts with the macroscopic universe).[16] When the system is observed, the wave function describing it is said to "collapse," which means it goes from inhabiting every possibility to choosing a single option.

By analogy, if you were a quantum mechanical system, your wave function might provide the probabilities of different locations you might be—for example, let's hypothetically say you have a 50 percent chance of being at home, a 20 percent chance of being in your car, a

10 percent chance of being at a coffee shop, a 10 percent chance of walking around your neighborhood, a 5 percent chance of sitting in a library, and a 5 percent chance of being at a friend's house. If you were a quantum system that had not yet been observed, you would both not be in any particular one of these locations *and* be in every one of those locations—until your wave function collapsed. There is a lot of weirdness here, but this is not a chapter on quantum mechanics, so hold on to those thoughts.

An important note is that the sum of those probabilities in the "where are you likely to be" analogy is equal to 1. What would it mean if the sum of the probabilities did not equal one? The only way we can really make sense of such a scenario is to concede that maybe we got the probabilities wrong, and we need to renormalize them so that their sum is once again equal to one. But that's not what I'm talking about—I mean what if the sum of the probabilities really truly does not equal one, meaning this lack of *unitarity* isn't something we can just normalize away? Having probabilities that don't add up to 1 is not allowed in quantum mechanics, and this is at the heart of the "principle of unitarity." This principle of unitarity is where the conservation of information comes from. In a nutshell, the sum of probabilities of every possible quantum state must be equal to one. If we lose quantum mechanical information—like by throwing a deck of cards into a black hole—this sum of probabilities is no longer one.

This apparent violation of the principle of unitarity comes into sharper focus when we think about quantum entanglement. If you haven't encountered quantum entanglement before, it is one of the most bizarre aspects of twentieth-century physics. Einstein hated it (and tried to disprove it until his death).

I find it easiest to explain how quantum entanglement works with (yet another) example. Let's envision we have some crazy physics experiment in the lab—there is a giant metal contraption with lots of knobs and dials in the middle of the room that occasionally makes clicking and whirring noises. The job of this machine is to create a pair of particles and send them off in different directions.

Because of the way this machine makes the particles, they are *entangled*—meaning that the property of one is related to the property of the other. In this case, we could imagine that one particle must be a triangle with the pointy side up, and the other must be a triangle with the pointy side down. But we don't know which triangle is which until we are able to measure one of them. The fancy machine goes ahead and creates this triangle pair and sends one member of the pair in each direction to be measured by separate machines waiting in the wings.

One particle makes it to the measuring device a tiny bit before the other, and we see that this triangle particle has its pointy side down. We *instantly* know that the other triangle particle has its pointy side up without even needing to go look (although we could). That all seems totally normal, at least insofar as crazy physics labs are "normal." Here is the thing—*before* we measured the first triangle particle, its wave function had not collapsed; it was in a hybrid state of both pointy side up *and* pointy side down (and neither pointy side up nor pointy side down). The instant we measure this triangle particle, its wave function collapses into one of these two states. That very same instant, the *entangled* triangle particle that went in the other direction must collapse into the other state, which—by virtue of happening instantly—took place faster than the speed of light. One modern theory in physics is that the very connection between entangled particles makes up the "fabric" of space-time—literally knitting together entangled particles via something like unfathomably tiny wormholes connecting them.[17] There are nights when I can't fall asleep because my brain gets stuck thinking about this concept and its implications—that space-time itself could be a manifestation of our connection to everything in the universe.

Einstein famously called this phenomenon between entangled particles "spooky action at a distance," and even devised a version of the above experiment (along with Boris Podolsky and Nathan Rosen, now called the "Einstein-Podolsky-Rosen experiment," or EPR) in an attempt to prove that quantum mechanics was incomplete, and thus advocating for "hidden variables" that predetermine the state a wave function will collapse into. The idea behind hidden variable theory is that there is some parameter that we are not aware of (aka "hidden") that tells the particles which state to collapse into, meaning the wave function collapse is not truly random at all. However, hidden variable theory was more or less put to rest by some clever math known as "Bell's theorem" for local hidden variables, which was subsequently experimentally tested. Proving Bell's theorem is now a classic problem for undergraduate physics majors, but I refuse to drag you through the logic and statistics of it here. If you are so inclined, it is easy to find explanations of Bell's theorem online (or in actual books, like I had back in the day when I was an undergraduate). I will note, however, that *non*-local hidden variables are not disallowed, which we will come to in Chapter 7.

Instead of sending one of these entangled triangle particles to a measuring device in the lab, what if we send it into a black hole? When the particles were created, each had a 50 percent chance of being pointy-side up or pointy-side down. Now that we have removed the information in the first particle's wave function from the universe, how do we get back to a sum of probabilities being equal to one? We have violated unitarity, and this is how we end up with an "information paradox."

As with entropy, the first solution to come to mind might be that the information still exists in the black hole beyond the event horizon; we just don't have access to it. After all, there is no law of physics that dictates that humans must have access to all information. This hidden-behind-the-horizon idea was a totally viable hypothesis, until Stephen Hawking pointed out that black holes can evaporate. Once a black hole evaporates, where does the information go?

As with many great mysteries, there are several working hypotheses that come at the information paradox from different directions, which range from mainstream to fantastical. Without the luxury of observations to test these hypotheses, I am a proponent of entertaining even the reasonably outlandish; deeply considering the various conjectures may well yield new insights and ideas, even if they turn out to be wrong. Unsolicited advice I often give my students (perhaps to the chagrin of their parents) is that if they always do splendidly well at everything, they probably aren't getting outside their comfort zone enough and they need to branch out. I feel the same way about experiments; if our hypotheses always turn out to be correct, we are probably not giving enough serious consideration to ideas off the beaten path.

In recent years there has been a flurry of activity that points to solving the information "paradox," which tends to be very technical. Many physicists seem fairly convinced that they have found a solution (and there are different solutions), but I'm not yet convinced that the case is closed for three primary reasons: (1) all of the existing possible solutions rely on (necessary) approximations because we don't yet have the mathematical tools to do the calculations exactly; (2) theorists continue to identify new hypotheses, which tells me we haven't completely explored the landscape yet; and (3) none of the hypotheses have been experimentally verified, and all we have to go on is consistency with other physics that we think we understand. I love theoretical physicists as much as the next person, but sometimes they benefit from observations to keep their confidence in check. I also have a bonus fourth reason, which applies beyond the topic at hand: we have a solid history of thinking we *know* things, only to find out later that we were wrong. This isn't a bad thing per se—this is how science is supposed to work as we gather new information. Rather, what nags at me is the arrogance in believing that we are right at any given time, which inhibits exploration of other possibilities—to me this is like badly pruning a plant and cutting off branches that might still bear fruit or help the plant to flourish. (Nope, definitely not speaking from personal experience.)

With that preamble, the possibilities at hand for solving the information "paradox" generally fall into three broad categories: first, information is actually lost, and through some mechanism this is allowed by physics—if information loss doesn't violate physics, we don't have a "paradox" to begin with (this option is generally dismissed out of hand because it would require completely turning physics inside out); second, the information isn't lost, but is stored somewhere, like the surface of the black hole where it can (in principle) be accessed by the outside world or be encoded in radiation; and third, something happens within the black hole that leaves a relic or imprint after the black hole evaporates.

Below are some examples of potential solutions that have been proposed. They range from mainstream to heterodox, but I would argue that since we don't yet know the answer, we should keep the wild cards on the table.

The first option is known as "black hole complementarity."[18] The idea here is that to an observer outside a black hole, they would never actually witness information falling in with a different, hapless observer crossing the event horizon into the maw of the black hole.

Sure, the observer falling into the black hole along with the information would argue the information went along with them, but since the two observers can never communicate, we don't have a contradiction (or so the argument goes). This proposal also includes a stretched horizon above the black hole, which would cause anything falling in to be heated and reradiated as Hawking radiation, which feeds directly into the firewall hypothesis mentioned earlier.

What if quantum mechanics saves the day? Black holes may or may not exist. Stephen Hawking gave a famous talk in 2004 in which he did an about-face on the information paradox, thereby conceding a bet—which resulted in his gifting a baseball encyclopedia "from which information can be recovered with ease."[19] Hawking argued that quantum uncertainty prohibits an outside observer (at an infinite distance from the black hole) from knowing whether there really is a black hole:

> Information is lost in topologically nontrivial metrics, like the eternal black hole. On the other hand, information is preserved in topologically trivial metrics. The confusion and paradox arose because people thought classically, in terms of a single topology for spacetime. It was either R4, or a black hole. But the Feynman sum over histories allows it to be both at once. One cannot tell which topology contributed the observation, any more than one can tell which slit the electron went through, in the two slits experiment.[20]

If black holes both *do* and *do not* exist, then we have a loophole for information. I realize that because Hawking said this, many people will be inclined to take it as the gospel truth. To be sure, he was a brilliant human, but even he had to make approximations in these calculations that we need to keep our eyes on.

The next option is the holographic principle (aka anti–de Sitter/conformal field theory correspondence, aka AdS/CFT). Sorry about the jargon, but if you want to crash a physics party, mention "AdS/CFT" at the door, and the host will assume you are a card-carrying member. This principle relies on string theory, which itself can be a bit polarizing in the community (we will talk about string theory in more detail in the chapter on other dimensions). For the record, I am currently inclined toward string theory, but withholding a smidge of confidence if for no other reason than theorists unhindered by actual observations make me uneasy.

Even before the Hawking talk in 2004, AdS/CFT was becoming a crowd favorite for solving the information paradox. There is way too much to unpack here, so I'm going to leave most of it in a suitcase you can choose to open later with a wealth of material online. In brief, AdS/CFT is duality between quantum field theory and general relativity that emerges from string theory and results in a holographic principle—which can encode information on the surface of a black hole.

In brief, the idea is that the full 3D information of an object can be stored on a 2D surface (i.e., a hologram)—this may seem like it ought to be impossible, but you may well have proof of this concept in your pocket right now; many credit cards use holograms, or if you happen to have a $100 bill around, you can check that for holographic behavior, too.

One circumstantial bit of evidence that supports the holographic principle is that theory suggests the entropy of black holes increases as the radius squared. If that behavior doesn't seem odd at first blush, think about it this way: one might expect the entropy of larger objects to be higher—in other words, one might expect entropy to increase with volume, and volume increases with radius *cubed*. However, the surface area of an object increases with radius squared—which is the same theoretical behavior of the entropy needed for black holes. This counterintuitive tidbit suggests that there is something special about the surface of the black hole that has to do with entropy. And this correspondence suggests that something like a holographic principle might be at play.

If information is stored on such a hologram, then it isn't lost from the universe and the principle of unitarity is safe. Phew.

The fourth option is the fuzzball hypothesis (FYI, this really is what they are called,[21] which as far as jargon goes is fairly friendly to a novitiate). These fuzzballs would effectively be a new state of matter even more extreme than neutron stars. This conjecture relies on AdS/CFT (and, in turn, on string theory, for better or worse). The basic idea in the fuzzball hypothesis is that at densities more extreme than those in neutron stars, the constituent neutrons effectively "melt" into the strings that make up their quarks. In turn, these newly liberated strings combine into longer structures, their tension decreases, and they can stretch out. When I envision the newly liberated strings expanding within fuzzballs, I find it tempting to almost anthropomorphize them, as if they have escaped from some confining cages of fundamental particles they were locked in while living outside of the black hole, but now they can flourish in the wild.

A remarkable feature of the fuzzball hypothesis is that it predicts black hole sizes that are exactly the same as we would calculate with the classical theory (i.e., a Schwarzschild radius). This is too extraordinary a coincidence to ignore and gives this hypothesis a bit of street cred. When we can't test things observationally, the best we can often do is look for consistency with other well-established physics, so the resulting predicted size scales of fuzzballs is important.

In the fuzzball conjecture, any infalling information gets encoded on the strings, which are in turn encoded on the fuzzy surface, which can then be encoded in Hawking radiation. And thus, we will have saved our universe from a paradox.

Our next option to avoid a paradox is that black holes actually have hair after all, just really soft hair. In Hawking's last paper (published posthumously), he and colleagues Malcolm Perry and Andrew Strominger tapped into a set of symmetries that are present in "asymptotically" flat space-time (which basically means space-time far away from a gravitational object like a black hole) known as "BMS symmetries" (after physicists Bondi, Metzner, and Sachs).[22] These special symmetries, called "supertranslation" symmetries, cause a halo of low-energy quantum excitations around the horizon of a black hole, which are referred to as "soft hair." These soft hairs can encode a memory of what has fallen in. To be sure, and by Hawking's own admission, this conjecture hasn't yet resolved the information paradox, but it does provide a fun new toy to play with in our models.

At this point the conjectures to avoid the information paradox get extra fun, starting with quantum wormholes. Two of the luminaries in this field, Juan Maldacena and Leonard Susskind, put forward a conjecture that uses quantum scale wormholes (or formally, "Einstein-Rosen bridges") that enable entangled particles to exchange information—across the black hole horizon, which they refer to as the "ER = EPR principle" (explicitly connecting Einstein-Rosen bridges with the Einstein-Podolsky-Rosen experiment).[23] In this case, not only is the information conserved, but there is another lovely side effect of this hypothesis—it would help us avoid black holes having those inconvenient firewalls.

Or, what if after a black hole evaporates it leaves some type of remnant that could encode the information? For example, the final quantum state left over after evaporation could be a unique vacuum state.[24] But at some point we do need to consider what might happen if quantum mechanics or string theory don't save the day.

In this case, we need to lean into what space-time itself might do. If gravity is left to its own devices, deep inside the event horizon the warping of space-time would be *extreme* (where by "extreme" I mean "infinite"). The question is whether this extreme warping leaves an imprint in the very fabric of space-time. The general idea in this hypothesis is that when the black hole is evaporating, once it reaches a quantum size scale (i.e., the Planck length), we are left with something like a quantum-sized topological knot in space-time.[25] These remnants have even been proposed as candidates for dark matter (which is a whole other mystery we will talk about soon).[26]

I've saved this possibility for last because I think it is by far the most fun to think about (and perhaps also the most controversial): What if the information goes *elsewhere*? By "elsewhere," I mean like *another universe*. As the fabric of space-time is stretched to extreme levels near the central singularity, it could well be that some form of inflation[27] kicks in and balloons the "internal" region out— potentially into an entirely new universe. The idea here is that the "inside" of a black hole can actually be larger than the outside horizon; if we deploy an embedding diagram, it might look something like this:

John Wheeler referred to this option as a "bag of gold," into which one could stuff a whole lot of information. Lee Smolin has postulated the creation of baby universes by this mechanism, and a fabulous debate has ensued.[28] A particular sticking point hinges on whether any of the other mechanisms for solving the information paradox above are correct; if no information is "lost" to the black hole, it is hard to imagine how one could seed a new universe. This also brings us back to the Big Bang and where the information in our own universe might have come from. Baby universes also bring us to the center of a black hole and whether we have a singularity.

The "Singularity" and Beyond

Within the event horizon of a black hole, our understanding of physics enters a hybrid state; we believe we have some knowledge and understanding based on an extrapolation of well-understood physics, but we also know that even our most well-grounded theories are likely to break down near the singularity. Within the event horizon, we have no cause to think that our understanding of physics will fail, at least not until we are considering realms deep inside the black hole. Sure, we expect a lot of extreme behavior—like time becoming imaginary, and space itself flowing toward the singularity faster than the speed of light, but there is nothing particularly special about the event horizon that disrupts the otherwise well-understood math and physics.

Of course, we can't empirically *test* the extrapolation of our understanding with real live black holes, which renders everything on the inside of the event horizon firmly in the borderlands of science. The dearth of empirical inquiry opportunities doesn't stop fearless physicists from generating hypotheses, though, some of which have aspects that may be at least partially tested outside black holes.

Because hypotheses about what happens within black holes are irritatingly unconstrained by observations, the only real assessment of their validity available to us is their consistency with the rest of physics. But given that we know we don't (yet) understand physics near the singularity, it is hard to be truly incompatible with potential new physics that may come into play in that realm.

Different possible solutions to the information paradox from the last section provide entirely different scenarios for what might happen inside the event horizon, and there may not even be a central singularity. Solutions that avoid a singularity are alluring because singularities are uncomfortable both mathematically and philosophically. After all, they are very sharp and pointy.

If there are central singularities in black holes, they are arguably the most extreme physical states in the universe. Most self-respecting physicists I know just stop talking about black holes at this point, and for good reason—we know we don't know what happens. Acknowledging that stupidity and bravery can look very similar, I'm going to keep going, in part because this is where things get *really* interesting—if I stopped writing every time we hit the boundary of what we know, the title of this book would need to be something more like *Things We Can Empirically Test About the Cosmos*. Besides, to quote once more the second law of the late Arthur C. Clarke, "The only way of discovering the limits of the possible is to venture a little way past them into the impossible."

Given the extreme conditions at or near the singularity, some fun possibilities emerge. Maybe quantum mechanics takes a stand and tells gravity to back off, and we avoid a singularity altogether.[29] Intuitively, you can see how this might play out: at the point where the mass within the black hole is compressed to quantum scales, Heisenberg's uncertainty principle may make a singularity in space-time impossible—both in space and time. In quantum mechanics, there are pairs of variables that are tied together—for example, position and momentum, and time and energy. The more precisely you know the value of one member of the pair, the less precisely it is possible to know the value of the other. In normal macroscopic life, this really doesn't affect us, which contributes to this phenomenon being nonintuitive (along with much of quantum mechanics). The upshot is that trying to precisely localize values of these variables at the singularity may end up generating a writhing and chaotic tangle. You could be justified in thinking of this as a new state of matter, and one with which we humans have no experience.

There could also be a wormhole (or, more properly an Einstein-Rosen bridge) that connects to another region in space-time. With a nod to the reality that we don't yet have a testable theory of quantum gravity and our understanding may well change with future advances, as far as we know at the moment, wormholes are allowed solutions in general relativity. This is not the same as saying wormholes exist, but it does mean that we can't say that they don't. Whether or not wormholes could be traversable (and used legitimately as plot devices in science fiction) is another matter altogether. On the face of it, gravitational theory suggests that wormholes would be *highly* unstable to collapse—which begs the questions of what would happen to you if you got stuck in such a collapse, and I feel confident that it would not be good. However, there are hypotheses on the table that could endow wormholes with stability—including the concept of negative pressure, which will play an outsized role in the chapter on dark energy.

A traversable wormhole would also require a *white hole* connected to the black hole if you wanted to end up somewhere else. You can think of a white hole as effectively the opposite of a black hole—instead of things being doomed to fall in at the event horizon, white holes expel light and matter and cannot be entered from the surrounding universe. One glaring issue with white holes is that we have never seen one—although some scientists have proposed that certain gamma ray bursts that have been recorded could be white hole events. Another intriguing possibility is that the Big Bang itself was a white hole event, which brings us to the next option.

It could be that a new universe is created. This is a legit hypothesis. In fact, our own universe could have been created this way. In a nutshell, once space-time becomes sufficiently stretched within the maw of a black hole, negative pressure takes over (which acts like negative mass). The space-time would radically inflate, and a whole new baby universe could pop into existence. This process would look a lot like what we know about the Big Bang, and the upshot is that we could very well be living inside of a black hole created in a parent universe. Physicist Lee Smolin took this to a new level and hypothesized that this process could give rise to "cosmological natural selection." In other words, universes that have properties that make them more likely to form black holes are more likely to have lots of offspring. If there is some level of inheritance of these physical property "genes" with a child universe, then the resulting children are more likely to provide grandchildren. And so on. This topic, once again, solidly intersects with other chapters in this book—including the Big Bang, where the laws of the universe come from, and whether the universe is fine-tuned.

Finally, we don't know what we don't know, and maybe what happens inside a black hole is something else entirely. Given that we are outside the realm of empirical inquiry here, it would be misguided to believe we have thought of all the possibilities. After all, there is a lot we don't understand about the universe, including how space-time behaves near the center of a black hole and the nature of time itself.

Interstition

With both the Big Bang and black holes, we were forced to come to terms with time being a full partner in the fabric of space-time. With the Big Bang, we had to question whether time itself had a beginning (and whether it could have an end) and grapple with the incongruence of causality in a temporally deprived domain. Black holes brought the malleability of time to the forefront—that time can get stretched out and slow down, and even become imaginary.

The nature of time strikes to the very core of infinite regression and how we think about the foundations of the universe. Our encounters in the previous chapters have pointed out the mysteries of time that are hiding in plain sight; time is truly an odd mixture of being simultaneously so familiar we don't even think about it and yet so completely foreign when we stop to consider what it is.

Since the nature of time keeps creeping into the chapters and trying to get our attention, I am going to give time its very own chapter. With this coming chapter we leave tangible solid ground behind.

7

What Is the Nature
of Time?

INDULGE ME FOR A MOMENT—TAKE A MINUTE OR TWO AND think about how you would define time.

Time is a slippery concept. We interact with it continuously throughout our lives, and much of physics is defined with respect to time, yet we don't understand what it truly is. I would argue that time is perhaps the single concept we simultaneously have the most exposure to and the least understanding of. It is ridiculously easy to go through life utterly taking time for granted as just part of the background.

One of the problems inherent to talking about time is that we are stuck in its flow and don't have a common vocabulary to talk about time from *outside* of time. For example, if we use an analogy of a river to talk about time, words like "flow" and "rate" make sense—we have a riverbank *outside* of the river as a reference point. If we want to discuss whether the rate that time passes might change, what is this rate with respect to? What is the "riverbank"? Clearly something like a rate of time passing per hour doesn't work. Since we are embedded in time, much like a fish in water (or so I am told, but not firsthand from a fish), it is hard for us to even properly perceive time for what it is. This is another one of the concepts that evolution—both biological and linguistic—has not properly prepared us for thinking or talking about.

The nature of time, like many possibly unanswerable questions, has been debated for millennia. For many generations, the debate centered on whether time is solely *relational* (meaning time only exists as an abstract concept relative to things that happen), or whether instead time is *absolute* (meaning time exists in and of itself, regardless of whether anything happens).

To illustrate the thinking of some renowned thinkers, here are some illustrative quotes that fall into one camp or the other:

"Time is the measure of change."—Aristotle

"Absolute, true, and mathematical time, in and of itself and its own nature, without reference to anything external, flows uniformly and by another name is called duration."—Newton

"Whether things run or stand still, whether we sleep or wake, time flows in its even tenor."—Barrow

"I hold [time] to be an order of coexistences, as time is an order of successions."—Leibniz

"The conceptions of time and space have been such that if everything in the Universe were taken away, if there were nothing left, there would still be left to man time and space."—Einstein

If even these intellectual giants can't agree, what gives us cause to think we can make a dent in this enigma? Well, for starters, we have a slight edge on our predecessors—the concept of time used to be entirely in the realm of philosophy, but modern physics has reframed the discussion and has something to say.

What is time? Why does it (appear to) have a direction? Does it actually flow (and if so, flow with respect to what)? What determines the rate of time? Does time have a shape? Could time have a beginning or end? We don't yet have the answers, but a few clues have shown up in modern physics.

Insights from Relativity

The concept of time was nicely ensconced in the domain of philosophy until special relativity walked into the room, followed shortly thereafter by its sibling named "general." Given the ability of the two relativity siblings to turn physics upside down, they are based on remarkably mundane and simple axioms: the laws of physics behave the same everywhere, acceleration and gravity are indistinguishable, and the speed of light doesn't care where you are or what you are doing.

I recall getting in a fight with a friend about this in elementary school—the friend was proudly telling me that if you were traveling at a constant speed, it wasn't possible to tell whether you were moving. I was utterly incredulous and refused to believe it. Surely when you are driving in a car, you can tell that you are moving! Indeed, you can tell that *something* is moving, but is it you or the outside world? If you have ever experienced being in a car stopped next to a large vehicle that filled your field of vision when that other vehicle started moving forward, you perhaps felt that unsettling sense that you were moving *backward*. That is everyday relativity in action. Of course, the effects of relativity get increasingly bizarre as we consider less prosaic situations.

Because the temporal effects of relativity are outside of our perception in the normal conditions of human experience (for which we should probably be grateful), this malleable nature of time is very easy to ignore, and equally difficult to intuit. The relativistic effects on time have, however, been verified over and over, on the ground, in airplanes, and in space.

My own children, who have been raised by two scientists as parents (much to their chagrin), have been known to invoke relativity as a loophole in taut situations. One particularly chaotic morning when I was dashing about, I needed to first drop off a cat at the vet and then rush the kids to their respective schools. Driving out of the parking space at the veterinarian's office, I backed right into a truck. A really big truck. A really big truck that had been sitting there since before we got to the vet. The truck was 100 percent fine. My little Prius not so much. My darling children—without missing a beat—immediately suggested that, in fact, because of relativity, I had not been moving, the truck had, and therefore it wasn't my fault. I do wonder whether anyone has tried to use this argument in court.

The relativity siblings have some foundational things to say about time that fly in the face of our normal daily perception. Perhaps settling the age-old debate on whether time is absolute or relational, relativity tells us that time is absolutely a thing in and of itself and woven together with space into the underlying substructure of the universe. But with a nod to the relational position on time, time is also absolutely not universal—every location in the fabric of space-time experiences time in its own way, which is where things get wild.

We need only think back to the chapter on black holes to see how these time shenanigans might play out; in general relativity the "speed" of time is contingent on exactly how much gravity an object is subjected to. To put a finer point on this, even the cells in your feet are aging at an (immeasurably) different rate than the cells in your head (unless you spend most of your time standing on your hands instead of your feet, in which case just invert what I said). If you happen to spend a lot of time on high-speed airplanes, special relativity tells us you will have experienced (imperceptibly) less time than people who stay on the ground using much slower modes of transportation. Sometimes I wig myself out a bit by realizing that my spouse and I are at different places in time, with one of us inevitably having experienced ever so slightly more or less of it. FYI, this is true of every person you interact with.

I want to pull out something from the previous paragraph and put it under a magnifying glass. This fact might seem trivial on the surface but speaks to a deeper underlying truth; objects *fall down* into gravitational potential wells. Big deal, right? I've witnessed my own children verify this since their early experiments intentionally dropping food from their high chairs. My cats, being normal cats, seem to find gravity infinitely fascinating, knocking objects off shelves and counters whenever possible. Simply noting that things *fall down* seems utterly banal. The key is making the connection to what this means about the rate of time for the object. Another way to say this is—unless acted on by another force, objects move toward the "slowest" time available to them. To my ear, this rings of something foundational about the universe and time. The lowest state of energy for an object involves experiencing the least (relative) time.

Another aspect of relativity that is particularly vexing is that there is no such thing as a universal "now"; granted, in our own little provincial corner of reality, it sure seems like there is a "now," but if we had the ability to perceive or measure extraordinarily small increments in time, we would find this not to be true. What "now" is for you is not the same as "now" for your best friend. An analogy I often use to visualize this is a loaf of raisin bread that you can slice at different angles. This loaf represents space-time (with two dimensions of space along the face of the bread and one of time along the length). Each slice represents a "now" at all locations in a two-dimensional slice of space for a particular observer. Your "now" will be at a slightly different angle than my "now." In other words, a raisin in my "now" slice might be your future. Or a raisin in my past slice, might be in your present.

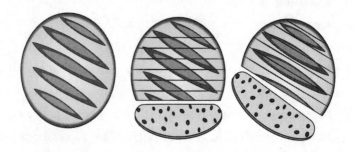

To be explicit about a key point from relativity—it is functionally legitimate to infer that the past "still" exists, and the future "already" exists. I hope you find that as utterly vertigo-inducing as I do. This line of thought leads us deeply into considering how far we can take this raisin-loaf analogy as we consider three potential options for time:

First, only a single slice exists—otherwise known as "presentism." "Now" exists, and "now" is all that exists. Relativity suggests this option is off the table because which angle slice of time gets privileged as being "real"?

Second, the loaf is built one slice at a time into the future. In other words, in some sense the past can be said to still exist, but the future not yet. As above, we encounter an issue with which angle slice gets to be at the front of the loaf. So, again, relativity suggests this can't work.

Third, the whole loaf exists. Relativity doesn't object to this, but now we have a new essential question: If the whole loaf exists, why do we find ourselves perceiving the slice of time that we do? And what does this mean for free will?

Insights from Quantum Mechanics

As the other pillar of modern physics, quantum mechanics also has something to say about the nature of time. But first, we need some quick background on how quantum mechanics works—just a couple key concepts. As with pretty much anything having to do with quantum mechanics, these concepts may not make sense, but at some point, you must accept that "common sense" and "reality" are frenemies.

The first core concept in the quantum world is the nature of uncertainty. In proper quantum mechanical style, this is not the kind of uncertainty you may be accustomed to in your daily life. For example, perhaps you're hosting a party and 15 people have said they will attend. But as a seasoned party host, you are well aware that a couple of the guests may show up with an unexpected +1, and conversely, a couple of the guests will get sick or have something last-minute come up and not be able to attend. As a result, you are expecting perhaps 15 ± 2 people, and you plan accordingly.

This is not how quantum uncertainty works. In quantum mechanics, the uncertainty doesn't come from *you* not knowing the answer, rather quantum mechanical uncertainty is rooted in a specific answer *not existing*. There is a world of difference between *you not knowing* precisely how many people will show up at your party, and the number of people who will show up fundamentally *being unknowable*—by anyone, ever—until the number of people is observed. Moreover—every possible number of attendees exists simultaneously, and reality doesn't decide what the final number is that will truly attend until your party starts, which makes planning a real beast.

An understanding of quantum uncertainty is inherently tied up in the nature of time; quantum mechanics tells us that the time of an event[1] can only be known with a certain level of precision, below which time only exists as a fuzzy concept without discrete values. This granular quantity of time—within which time does not exist in the way we know and love—is the Planck time, or 5.39×10^{-44} seconds (which we often round up to 10^{-43} seconds, because really at this scale, the level of precision makes no difference to our ability to understand that this is small). Because the normal concept of time doesn't make sense on scales less than Planck time, this scale could be the fundamental building block of time. In other words, maybe time is not actually continuous. Instead, quantum mechanics suggests that time may come in fuzzy little portions called "chronons." You can sort of think of these as atoms of time.

If time isn't continuous, but rather comes in discrete units, would we know? Briefly, no. Human perception is limited to less than one hundred frames per second (the so-called "flicker rate"), and videos with a frame rate of less than twenty-four will appear choppy, and this flicker rate is fundamentally limited by our physiology. Of course, our physiology is somewhat limited. By contrast, the fastest current clocks on Earth measure intervals of time down to an astounding $\sim 10^{-18}$ seconds. If the underlying granularity of time is anywhere close to the Planck time, we are still over twenty orders of magnitude away from accessing time intervals at that scale.

The second essential element of quantum we need to bring to bear on this topic is that characteristics of things are described by "wave functions" (almost always denoted by the Greek letter psi Ψ). "Wave function" is kind of an odd term—why are we describing things (particles, atoms, cats) with something having to do with *waves*? Good question. Because it turns out everything is fundamentally a wave, even if we don't perceive it as such. Hopefully, at this point in the book, you are getting accustomed to your perception not being the gold standard. Not only is everything (at the quantum level) described by waves, but these waves only encode the *probabilities* of different characteristics, not to be confused with actual specific characteristics. The actual characteristics of things are not determined until the thing is observed (i.e., interacts with the macroscopic universe), at which point the wave function "collapses" to a single discrete characteristic. *Why* this happens and *how* it works are among the great mysteries in modern physics.

I want to pause and point out something interesting about the Schrödinger equation, which is the equation at the heart of quantum mechanics and governs wave function behavior.

$$i\hbar \frac{d}{dt}|\Psi(t)\rangle = \hat{H}|\Psi(t)\rangle$$

To the uninitiated, this equation may look like it could cause bodily harm. But I want to focus on just the single letter i at the beginning. This i tells us that imaginary numbers are involved. My experience (backed by my children's and students' experiences) is that imaginary numbers are typically glossed over quickly in school as "simply what you get when you take the square root of a negative number." One way to think about imaginary numbers is that they are orthogonal to real numbers in a complex number plane (with real numbers forming the x-axis and imaginary numbers forming the y-axis).

It would be one thing if imaginary (or complex) numbers were just this nifty esoteric math trick, but instead imaginary numbers are entwined with many of our most essential physical concepts. In this vein, I can't resist mentioning Euler's equation, which permeates physics:

$$e^{i\pi} + 1 = 0$$

Insofar as mathematics can induce mystical experiences, Euler's equation is at the top of the candidate list. Not only does this equation include *i*, but it literally relates the most fundamental numbers in the universe in one incredibly economical equation with *nothing* wasted—no random values or components. If I were to get a tattoo of an equation, this would be the one. It is very hard to see this equation and know the truth of it without an acute sense that something else is going on behind the curtain.

When I was first introduced to the concept of imaginary numbers as a child, I thought they were super intriguing (don't you agree?) but could not really get much of a sense of their "reality" from my teachers. If you take a square root of a negative number, you get an imaginary number, and that is the end of the story. Later I learned as a physicist that we throw around imaginary numbers all the time in various equations. For the most part I think we view them as useful tools that help with mathematical descriptions, but we don't often pause to consider what it means that we need them. Now, firmly in middle age and not caring about sounding silly, I've come to think of imaginary numbers as signaling that there is a concept hovering just outside our standard view of reality, and we should sit up and take notice. So the fact that the behavior of wave functions is tied to imaginary numbers seems appropriately weird for quantum mechanics. OK, now back to Schrödinger's equation and quantum mechanics.

Given that the future state of a thing is not determined until its wave function collapses, it is natural to think wave function collapse might have something to do with time. For example, one could envision that time takes a step forward with every wave function collapse, and the future is built one wave function collapse at a time. But the thing about wave function collapses is that they are (or at least appear) *random*, and this has enormous implications for the nature of reality, depending on the interpretation of what is happening under the hood.

There is a collection of wave function collapse interpretations, each of which has the power to change our view of reality in a different way. The interpretations mainly vary in whether they are deterministic (versus random) or require quantum nonlocality (which means quantum connections can influence what happens faster than light). Going through them all would get tedious (at least for me, and I assume for you), so I will just give you a sense of a few of the most discussed.

The Copenhagen interpretation is perhaps the easiest to explain: when a wave function collapses, the resulting characteristic that becomes reality is truly random. This may seem well and good if you are OK with nature being fundamentally nondeterministic, which doesn't feel super comfy to me; I don't like there not being a reason "why" something happens. Aside from my personal comfort level (which is not known to be a good judge of reality), there is a tiny bit of tension between this and the idea that the full loaf of raisin bread already exists. But we already know that quantum mechanics and relativity don't get along, so maybe we shouldn't be surprised.

In contrast, in the many-worlds interpretation, the collapse of the wave function isn't random. Instead, every possible wave function collapse happens, and with each outcome a new branch of the universe is spawned off. In this case, there are versions of the universe out there in which everything possible has happened, both good and bad. The general discomfort with this interpretation is that we end up with a very large (probably infinite) number of branched universes. In which case, why do we find our consciousness in this singular branch? I often think about the many-worlds interpretation when I'm out hiking in the woods and must choose which way to go at a fork in the trail and wonder whether my choice will create a new branch of the universe. Go ahead and think about this every time you decide what socks to wear.

We neatly get around the bits of discomfort that appear above with the transactional interpretation of wave function collapse—by invoking "retrocausation." In other words, the future affects the past. In this way, the wave function collapse isn't random, but instead depends on information from the future propagating backward in time; since we aren't aware of this information, the result appears random. This interpretation also means that, at least in principle, the universe is perfectly reversible, with no messy stochasticity. And, as a bonus, no pesky branches of the universe in which you decided that one really terrible idea was worth doing. The flip side is that the possibility of free will goes out the window.

Finally, I want to mention the "pilot wave" theory, which I don't think gets enough attention. This interpretation was originally postulated by Louis de Broglie in the 1920s, but then abandoned for decades until David Bohm picked it up in the '50s. In this version of quantum mechanics, the characteristics of particles actually exist (regardless of whether they are observed), but every particle has an associated wave (the pilot wave) that tells it how to behave. You may be thinking that this sounds a lot like the "hidden variables" that Bell's theorem rules out, so how does pilot wave theory get around this? The catch is that the hidden variables in pilot wave theory are *non*-local; the pilot wave extends throughout space and—instantaneously (e.g., faster than light)—affects the behavior of entangled particles. To be sure, there is still weirdness here, just different weirdness than the other interpretations of quantum mechanics.

There are many more interpretations of wave function collapse out there, and one could (and many have) written entire books on them, so forgive me for this superficial treatment. Hopefully, just with these three interpretations alone, you have a sense of how profoundly our sense of time (and reality in general) might depend on the underlying mechanics of the quantum world.

The Direction of Time

I have yet to meet someone who doesn't perceive time as being asymmetric—only "flowing" in one direction. Why time appears to have a preferred direction and why we don't have any choice about how we move in time is perplexing. In fact, why the dimension of time seems fundamentally different than the three dimensions of space is one of the fundamental questions at play.

There is no doubt that our notion of time is intimately linked to causality, with a cause leading to an effect and not the other way around. This asymmetry of time seems perfectly natural in everyday life but is startlingly weird in physics; classical physics, relativity, and quantum mechanics are all time symmetric—equally valid run forward or backward in time. As a familiar example, think about a billiards table: if two balls collide and bounce off each other in different directions, you could reverse that process in time, and it would be totally normal. Yes, of course, I am ignoring friction, because friction makes everything more complicated. But the basic principle still holds—it is just that it is playing out at a microscopic level with friction. We can take this macroscopic example and shrink it down to the scale of interactions between particles, and the same reversibility holds.

Perhaps less familiar are Maxwell's equations, which (among other things) govern the propagation of light. These equations are so fundamental to physics that I know people who have them tattooed on themselves.[2] When physics students are dutifully doing their homework and using Maxwell's equations to solve for something or other, there are two possible solutions (think of this like the square root of 1, with both 1 and −1 being equally valid): one solution is a wave that goes forward in time (technically called "retarded"), as we would normally expect it to do. But the other solution goes *backward* in time (technically called "advanced"). Often, we just ignore the solution that goes backward in time as "unphysical," but why? There is no physical reason to rule this possibility out . . . other than that we don't observe it. But to be fair, it isn't clear how we *could* observe it. The fact that time has an arrow that seems to point forward in time is perplexing.

Perhaps this is not your first arrow-of-time rodeo, and you are thinking to yourself something like, "I know the answer—entropy!" If so, you are in very good company. My experience is that "entropy" is by far the most preferred answer among physicists to explain the arrow of time, and for good reasons. As it happens, I hold a contrary opinion.

Before I explain why I disagree, let's look at what it is about entropy that makes it a compelling cause for explaining the arrow of time. The crux of this argument is that the second law of thermodynamics doesn't mince words—entropy can only stay the same or increase in time in a closed system. This is in markedly solid agreement with our lived experiences—you can knock over a glass and spill milk, but I suspect you have never experienced spilled milk reversing itself back into a glass, which then stood back up. If you watch some time-reverse videos (which are available aplenty online and seem to be a favorite format for bored high school students), the only aspects that look especially whacky in reverse are those that blatantly violate the second law of thermodynamics—like the spilled milk reorganizing back into the glass. The crazy thing is, if every microscopic particle exactly reversed its action at the same time, this reverse action is what the milk would do. We just don't generally observe this because all the particles involved having the inverse action at the same time is unfathomably unlikely to happen. Everything else in a reverse-time video that doesn't hugely increase entropy looks normal-ish. Entropy is the singular area in which we overtly experience an apparently preferred direction in time.

All of this is quite true, but I would exercise caution. Do you remember reading about correlation and causation in the chapter on epistemology? The relationship between entropy and time is a prime example. Both entropy and time appear to go in the same direction, but that does not require that one *causes* the other.

Entropy is fundamentally a statistical statement. If we think back to the deck of cards we used in the last chapter, we found (not surprisingly) that with each successive shuffle, the entropy of the deck increased; the shuffled deck takes more statements to describe, which equates to more disorder. The reason the entropy increased was that the order of the cards in the crisp fresh deck was only one of a many possible states the deck *could* be in. As we shuffle, the order of the deck will wander through all possible orders. In fact, a standard deck of cards has 52! (here, the exclamation point is shorthand for "factorial" and not that I am just really excited about the number 52; mathematically 52 "factorial" = $52 \times 51 \times 50 \times 49 \times \ldots$) unique possible card orders. If you haven't plugged this into your calculator, you might be surprised by the resulting number—surely you are thinking there are bound to be a lot of possible card orders, but were you thinking it would be as high as 8×10^{67}? To put this number in perspective, there have "only" been about 4×10^{17} seconds since the Big Bang, so even if you had shuffled a deck of cards every single second since then, you would not have gotten to every order. By far. The next time you shuffle a deck of cards you might be mindful that the *particular* order your deck is in may well never have existed in the history of the universe.

You might consider the particular order of your shuffled cards to be especially important for some reason and want to shuffle back to it; however, it is just as unlikely that you will ever get that particular card order again by shuffling as it is that you will shuffle back to the initial order in the pristine deck (although, after shuffling ~10^{68} times, I am not sure there would be much of a deck left to be shuffled). However unlikely, it is *possible*—which brings us to the Poincaré recurrence theorem: given enough time, any system that randomly moves between possible states will return to its initial state. For a full deck of 52 cards this would take a *long time*, but what if we only use the aces—meaning we have 4 cards. Calculating 4! doesn't even require a calculator. With some efficient shuffling, you could revisit that state in just a couple minutes.

The metapoint of invoking the Poincaré recurrence theorem is that entropy does not *always* increase—it can and does decrease on small scales all the time. If entropy were causally connected to the arrow of time, that would require there to be micro–time reversals happening all around us every day and time would only move forward in a statical sense. I suppose we can't rule out that these micro–time reversals are happening, but I find it much more likely that entropy is merely correlated and not causally connected with the arrow of time. As time goes forward, things in the universe continue a statistically random walk through all possible states, but some of those states are lower entropy than others.

There is one additional intriguing tidbit I want to add into the mix of our discussion on time, which may mean nothing, or it may be a clue about something more fundamental. As an undergraduate physics major, when I first learned about these things called "Feynman diagrams" (after the late physicist Richard Feynman), they utterly captivated me. If I hadn't been hooked on physics before learning about them, they would have done the trick. These diagrams are a fantastically elegant way to conceptualize particle interactions. They made *way* more intuitive sense to me than endless equations describing the same phenomenon. At a fundamental level, these diagrams are just a two-dimensional plane of space and time that illustrates how particles interact. That is useful, but not necessarily mind-blowing.

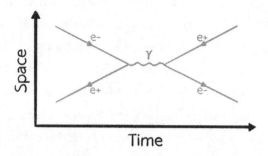

I sat up in my seat and became riveted when my professor told the class that when using these diagrams, antiparticles behave exactly like their normal particle counterparts—but going *backward* in time. For example, in a Feynman diagram, a positron is equivalent to an electron going backward in time (and vice versa). Moreover, when we talk about particle-antiparticle annihilation, this is functionally equivalent to a particle going forward in time meeting up with one of its alter egos going backward in time.

At the time, I also happened to be taking a separate physics class for which our weekly homework involved coming up with three novel ideas involving physics. I don't know how hard that might sound to you, but I promise it was both nontrivial and one of the best creative thinking exercises I've ever encountered in a physics class. The week we learned about Feynman diagrams, I wrote up a whole (naive undergraduate) theory about how the Big Bang could have been the result of a collision of a universe going forward and with a universe going backward in time, and I even worked out how this might also explain the matter-antimatter symmetry (which we will come to shortly). I'm now embarrassed by my (literally) sophomoric effort, but not too embarrassed to cop to it as a demonstration of how taken I was with Feynman diagrams. My homework came back with the comment "Interesting idea. Needs to be thought out more," which was both disheartening and fair.

To be clear, I am *not* saying that positrons are in fact electrons going back in time. I can almost hear a legion of dyed-in-the-wool physicists groaning that I even brought this topic up—this feature of Feynman diagrams is almost universally dismissed as a "neat trick" with no physical meaning (akin to advanced waves and Maxwell's equations, or imaginary numbers and Euler's equation). However, there is also not clearly an empirical way to test the idea that positrons are electrons going back in time, as they are functionally the same. My philosophical position is that when we are asked to dismiss solutions as "unphysical" that are otherwise legitimate, we should take a very careful look at our justification for saying they are "unphysical." We may well be cutting off promising avenues of thought before they even take root. Our human brains are limited enough—let's not make it even more difficult to come up with new ideas by convincing ourselves that things are impossible just because they seem unlikely.

The apparent asymmetry of time is fundamentally perplexing. Symmetry is a fundamental part of modern physics, and many physicists have an almost religious belief in the foundational existence of symmetry in the laws of nature (I would count myself among them). In physics, what "symmetry" refers to is a thing being identical after you have done something to it. Except, in physics, instead of "thing" we say "system," instead of "identical" we say "invariant," and instead of "done something to it" we say "it has undergone a transformation." So, restating this concept of symmetry in more physics-y jargon (you can break this out during small talk at your next gathering), symmetry refers to a system being invariant under a certain transformation. Transformations that often come up in the context of symmetry are reflection, rotation, and translation, but the list goes on and becomes increasingly abstract.

One of the great theories of the early 1900s came from Emmy Noether,[3] and it directly connects underlying physical symmetries to conservation laws. As a concrete example, let's take angular momentum. Lots of people have had *conservation of angular momentum* drilled into them at some point, so hopefully this concept is somewhat familiar to you, and you've seen (or you can imagine) an ice skater speeding up or slowing down their spin by moving their arms in or out. Conservation of angular momentum is tied to underlying rotational symmetry in physics—to put it bluntly, the laws of physics apply the same, regardless of how an object is oriented in space. We can do the same thing with conservation of normal (nonangular) momentum (which is connected to translation symmetry in space—moving an object linearly forward or backward along a dimension), or conservation of energy (which is connected to translation symmetry in time).

There are three particular types of symmetry that get entwined in physics at a really fundamental level: charge (C), parity (P), and time (T). Taking them each in turn: C-symmetry means that if you could replace every particle in the universe by the same particle with the opposite charge, the universe would behave the same. We can think of a classic atom as an example; if the proton in the nucleus had a negative charge and the electron had a positive charge, the atom would have the same properties. P-symmetry is basically just a reflection symmetry—if you take an object and reflect it along each of the three spatial axes, it would be identical (*P*-symmetry essentially filters for chirality, meaning whether something has a handedness to it). An everyday example of *P*-symmetry being broken is a clock; if you look at the face of a clock in a mirror, the hands will move counter-clockwise because clocks have a handedness to them. *T*-symmetry, as you might expect, means a system behaves the same whether you run it forward or backward in time—like the balls on a billiards table (ignoring friction . . .). Taken together, *C*, *P*, and *T* appear to be required to be symmetric *as a group*, which was a groundbreaking result in the 1950s.

You may well be wondering what it means for *C*, *P*, and *T* to be required to have a combined symmetry (referred to as "CPT symmetry"), which is not casually intuitive. This required CPT symmetry means that if I have a normal cat, moving forward in time with the rest of us, it would be identical to a cat made of antiparticles moving backward in time and reflected in a mirror. This required CPT symmetry is so important that, if we are wrong about this, much of our fundamental understanding of physics goes out the window. Which is not to say that we aren't capable of throwing physics out the window (I tried many times as a student) if we have no other choice.

We are getting to something essential and mysterious about the nature of time, and no one said this would be easy, so stay with me just a little longer. If we could talk about this over coffee with a napkin to write on, that would be great. But for the moment we are stuck in this book. The upshot: the arrow of time appears to be related to CP symmetry violations, which are in turn related to antimatter, which is itself related back to the arrow of time. So, it seems like we are onto something, we just don't know what.

Now we just need to remember how to multiply by 1 and −1, and we are home free. Let's say that if one of these three symmetries (*C*, *P*, or T) is symmetric, we assign it a value of +1. On the other hand, if it is asymmetric, we assign it a value of −1. For CPT symmetry to hold, the following must be true:

$$C \times P \times T = 1$$

There are, of course, different ways you can multiply +1 and −1 using these three variables and still end up with +1. For decades everything in physics was moving along and continually finding that $C \times P = 1$ (shockingly called "CP symmetry"), which worked well for interactions involving the strong force or electromagnetism. If $C \times P = 1$, then T also had to = 1.

Then, in 1964 along came this subatomic particle called a "kaon" (or "k-meson") and said, "Look what I can do!" And it had a partner in crime—the weak force, which I have found to be the most challenging of the four known forces to get a grasp of (hold tight for the chapter on the laws of nature—or don't and go ahead and read that now if you want). The weak force seems like this shy wallflower of the forces, coming across as unassuming and timid. To be fair, the weak force *is* weak (though not remotely as weak as gravity), and therefore trickier to study than the strong force or electromagnetism. Experiments got more sophisticated, and one day the weak force came out of its shell to show the world its talent; with the help of the weak force, the decay of the kaon violated CP symmetry. If CP symmetry has been violated ($C \times P = -1$), and because we need $C \times P \times T = 1$, then T must be −1.

If those last few paragraphs read along the lines of "Blah, blah, blah," here is the point: kaon decay was the first time in physics that we had ever found anything that appeared to be fundamentally asymmetric in time. In the years since the kaon had its coming-out party, we have found a few other interactions that have CP violation (and are therefore asymmetric in time), and *they all involve the weak force.*

As far as we can tell, the weak force is the only thing in physics that seems to know and care about the arrow of time. And we don't know why. If you have gone through your life blissfully ignoring the weak force as some esoteric force that only physicists care about, maybe it will pique your interest now.

It turns out that all of this is related to the fact that you exist, which you are presumably somewhat aware of.[4] Your existence would not be possible without an asymmetry between matter and antimatter in the universe. Given the extent to which physicists adore symmetry in the universe, this crucial asymmetry between matter and antimatter is particularly vexing. From both our theories and experiments, we unfailingly find that particles and antiparticles are created in equal numbers—as required by conservation of charge. So, this asymmetry really peeves us.

In the very early universe, matter and antimatter went through an intense period of mutual annihilation, turning the vast majority of their mass-energy into just energy. However, given that you are here reading this book, apparently the mutual annihilation wasn't quite complete. For reasons we don't (yet) understand, after all the destruction was done, something like 1/1,000,000,000 normal matter particles survived. If they hadn't, we wouldn't be here wondering about this asymmetry.

You might be wondering whether there could just be pockets of antimatter somewhere in the universe that we don't interact with. There *could* be, but they would have to be outside of our horizon (which doesn't rule them out). Even in "empty" space, there is on average about one atom in each cubic meter, so there really isn't likely to be a buffer of truly empty space between regions of matter and antimatter. As a result, the boundaries between these regions would be generating significant radiation from particles annihilating on the edges, and we haven't detected anything like this. Moreover, for the pockets-of-antimatter hypothesis to work, we would also have to explain—statistically—how matter and antimatter were not mixed at a level requiring annihilation. Given that particle-antiparticle pairs are created *together and are attracted to each other*, it is very hard to explain how clumps larger than galaxies could become segregated.

Perhaps the most promising hypothesis to explain the fact that the universe is dominated by matter over antimatter is CP violation. In cases of CP violation, we see that—for reasons we don't understand—mesons (like the kaon) have a very slight variation in how they decay that require $C \times P = -1$. Current experiments being run at the Large Hadron Collider are pushing the envelope on this, and perhaps this section of the book will need to be revised in the coming years.

Topology of Time

While time apparently having a direction feels natural from our lived experience, why time appears to be structurally different from space is a fundamental question in modern physics. Not only does there seem to be only a single dimension of time, but it also appears to only go in one direction, not offering us any choice about whether or not that is the direction we want to go. Because we seem to just be along for the ride as far as time goes, it might seem bizarre to even consider whether time has a "shape."

Insofar as most people even have cause to think about time at all,[5] I think there is a common conventional view based on lived experience. If I were to define a "common sense" view of time, it would be something like: Time is infinite (at least into the future).

Time is one-dimensional. Time only has a single branch. Time has a constant rate. Time goes in a straight line and doesn't loop around. Time is continuous. Time is relational and doesn't exist as a "thing."

I will be the first to admit that this "common sense" view of time is well aligned with my own perception. I've just learned not to trust my perception. How would we know if any of these statements were not true? Since we are embedded in time and don't seem to have much of a choice about that, it is very hard for us to get a solid grounding here. Nevertheless, let's take these statements one at a time and give them a proper inspection.

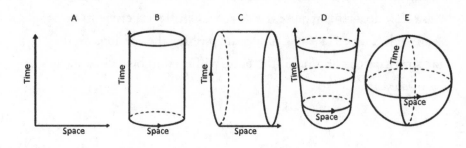

Does time have to be infinite? In a standard view of time and space, reinforced by countless physics and math diagrams, we commonly envision time as being orthogonal to space and unidimensional (as in Panel A).[6] Given the lack of constraints available to us empirically, there are literally an infinite number of topologies that time might have. For the sake of playing around with a couple toy ideas, let's explore a few different geometries as a thought experiment.

If the universe is finite in space, but infinite in time, the universe could have a shape in space-time similar to that shown in Panel B—if you kept flying your spaceship far enough, you could, in principle, loop back around to the same location in space (although the path for this in space-time would look like a spiral on this cylindrical figure, because you would also be moving up in time while looping around). The Panel B version of space-time is something that astronomers commonly envision when thinking about the universe, and we are not uncomfortable with the idea that the universe could possibly be finite in the spatial dimensions.

Now let's flip the script. If space could be finite, why not time? What if time were finite and space were infinite? This scenario would result in something like Panel C. Curiously, it seems easier for people (or at least for my students) to be comfortable with the idea of time having a beginning than an ending, which speaks to how differently we perceive time and space. Panel D is a version of the Hawking-Hartle no-boundary universe from the chapter on the Big Bang (you may want to flip back to that chapter for a refresher). In the "no-boundary" universe, the dimension of time becomes increasingly space-like the closer we get to a time $= 0$, and thus there is no "before" time $= 0$ as a result of the coordinate system.

Finally, the last permutation of these options: What if space and time were *both* finite? Then we might end up with something like Panel E. The situation in a universe like Panel E is interesting, because whether we view something as happening along the direction of time or space depends entirely on how the coordinate grid (latitude and longitude) is oriented; the symmetries embedded in this topology render the choice of coordinate somewhat arbitrary unless there is an anchor point—for example, because the Earth rotates, a practical coordinate grid is imprinted, and we get to have nice things like a north pole, south pole, and an equator.

To be clear, I am absolutely not advocating for or against any of these geometries. Rather, I think it is an interesting mental exercise to consider what it would mean if, for example, time were finite.

Does time have a single dimension? Next on the list of a "common sense" view of time is that it is one-dimensional. If time were two (or more) dimensions, you could—for example—turn "left" in time. One can imagine different incarnations of dimensionality in our universe, besides the 3 + 1 of space and time that we find agreeable. As a bit of a spoiler, the fact that we appear to have only a single dimension of time is probably critical for having a universe hospitable for life, as we will see in the chapter on fine-tuning. In a nutshell, if we had more than one dimension of time while also having more than one dimension of space, we end up with a mathematical situation that is probably not too good for life. In a scenario where you have multiples of one type of dimension (in our case, space), and also more than one of another type of dimension (in this case, time), you end up with what is called an "ultrahyperbolic equation." Ultrahyperbolic equations are pernicious and do not have well-defined stable solutions. If our universe had such dimensionality, nothing would be predictable, making it pure chaos at a fundamental level. If there is another dimension of time, it is thankfully hiding away somewhere and not destroying the order of the universe.

Could there be different branches of time? I don't see any reason this would be disallowed; in fact, the many-worlds interpretation of quantum mechanics leads to a branching of the universe into different timelines. In this case, it is interesting to consider why you find your consciousness in this particular branch and not some other. However, I also don't know how we would know if this topological feature of time were credible. To test this, we might need to find a way to get out of our own timeline, which for the moment seems rather impossible.

Does time have a constant rate? Special and general relativity definitively answer no to this question. But we can take this question to a level deeper—even if you are sitting still in the same place with the same gravitational field, is it possible that the rate of time universally slows down or speeds up? The most annoying part of this question for me is that it isn't clear what a "rate" of time would be with respect to. For example, if you were watching a toy boat float on a flowing river, you could measure the speed of river flowing with respect to the riverbank and end up with units of something like miles per hour. Since we seem to be trapped in the flow of time with no temporal riverbanks in sight, there is no obvious reference. For all we know, the general flow of time in the universe could be slowing down or speeding up.

If the general flow of time were to change for the entire universe, you might justifiably ask if this would matter (indeed, I have had students ask this in class). In terms of our own physical existence or observable consequences, I am not sure that it does matter in any practical way. However, as someone who cares about the fundamental nature of reality and the limits of our knowledge, I find it deeply unsettling to have so little understanding. Also, I'm just plain curious to know.

In terms of understanding the nature of time, our progress is mixed. Thanks to relativity, we do have a solid sense that time exists in and of itself, which is more than we knew a couple centuries ago. On the other hand, the quantum nature of time opens a line of questions that we didn't even know we didn't know a couple centuries ago. To be honest, I'm not even sure I can wrap my head around living in a world in which we understood the true nature of time and were presumably also able to manipulate it. In this case, Temporal Ethics would need to become a standard course in college curriculums.

For the moment, we are left needing to accept time and space as entangled dimensions that—for reasons we don't understand—seem to manifest themselves differently. Are there other dimensions that could be manifested still differently?

Interstition

Thinking of the dimensions of space as a sort of fabric is already removed from our lived experience, but adding time to this framework brings us face-to-face with the abstract idiosyncrasies of dimensions. These dimensions of space-time are the arena in which everything in our universe seems to play out, yet we understand very little about them.

The role of the dimensionality of our universe is entwined across the topics in this book. We have had to consider whether the dimensions of space and time might have come into existence with the creation of our universe. We have seen them behaving strangely near black holes, and we also have seen that hyperdimensional wormholes are not only not disallowed but may be responsible for the fabric of space-time itself. Even the laws of nature may be dependent on whether there are other dimensions (both big and small).

We seem to be restricted to going about our business in three dimensions of space and a singular dimension of time, but that doesn't mean there couldn't be other dimensions to which we don't have ready access. Of course, probing dimensions that appear to be inaccessible is a particular challenge for scientists, who, as a rule, are fond of empirical inquiry. However, it turns out there are hints of other dimensions in our theories, and as we think more creatively and abstractly, we are coming up with ideas that could help us explore beyond the veil of space-time.

8

Are There Hidden Dimensions?

You may have noticed that the universe we live in appears to have four dimensions—3 of space and 1 of time (which we often write in shorthand as 3 + 1 to indicate there are two types of dimensions). It is easy to take this dimensionality for granted, but this precise arrangement of operative dimensions turns out to be essential for life to emerge in the universe (more on this in the chapter on fine-tuning).

It could be that there are only 3 + 1 dimensions, in which case we might ask *why* 3 + 1? Could there have been a different number of dimensions? Similarly, we might ask whether there could be different *types* of dimensions, which are neither spatial nor temporal, but something else entirely. A different number or type of dimensions is challenging to intuit, if for no other reason than we didn't evolve to do so. Somewhere in our evolution it was important to be able to perceive and construct mental maps of our familiar 3 + 1 dimensions, which enables us to do beneficial things like move around and notice things happening. Given that these 4 familiar dimensions are the extent of the reality that we have ever perceived, it is natural to take these 4 dimensions (and only these 4) as self-evident—one of those axioms that seems so obvious we might go through life not even realizing we take this as given.

Given the limits of our experience and perception, the question of whether other dimensions can or do exist deserves a closer inspection. If other dimensions do exist, we might not perceive them for different reasons, depending on the characteristics of such dimensions. For example, hidden dimensions might be "compactified"— which means curled up on tiny scales that we can't yet probe, or they might not be coupled to the forces in our 3 + 1 domain.

If the forces we interact with don't extend beyond the familiar 3 + 1 dimensions, it is hard to fathom a way to probe whether such extra dimensions exist—the experiments and devices we can contrive generally implicitly depend on the known forces and their interactions. A positivist might argue that it doesn't matter if other dimensions exist if they have no influence on their universe. This position is hard to argue with, but that doesn't stop me from arguing—in particular, if we *assume* that either there are no other dimensions or we can't interact with them, this situation becomes a bit self-fulfilling.

Introduction to Dimensions

Let's clarify what we mean when we're talking about dimensions in this context. The essence of what a dimensionality means is *the number of coordinates necessary to pinpoint where something is*, which in the apparent dimensionality of our universe means *in space and time*. If something has zero dimensions (which corresponds to a point) then no coordinates are necessary; the thing in zero dimensions has no flexibility in where (or when) it might be, so providing coordinates is superfluous. The underlying concept with dimensions is that they are orthogonal to each other—the properties of each axis are independent of the others.

We can think about going from a lower dimensional object to a higher dimensional object by shifting (formally, we call this "translating") the lower dimensional object, connecting the vertices, and projecting onto a 2D surface—conveniently like this page of the book.

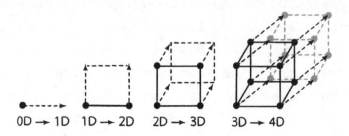

$0D \rightarrow 1D \quad 1D \rightarrow 2D \quad 2D \rightarrow 3D \quad 3D \rightarrow 4D$

In principle, you can keep going ad nauseam, but I don't have the patience to go beyond 4D in this scheme—there are way too many vertices to keep straight on a 2D sheet of paper. Heck, I have trouble even drawing a 4D cube freehand. At least with 4D, you can tell your brain to think about this extra dimension like it might consider time, and it isn't quite so hard to wrap our heads around—at one moment the 3D cube is in one position, and a moment later it has shifted to the other position.

When we are considering physical objects, it might be challenging to intuit how orthogonal dimensions would behave beyond the familiar 3 + 1. However, you use multidimensional constructs in your daily life, perhaps without even realizing it. So instead of just thinking about shapes and physical objects, we can use an example that is near and dear to my heart: food. Food can be described along many axes that are *independent*: sweetness, saltiness, spiciness, crunchiness, temperature, caloric value, color, and so on. When deciding what you might be in the mood for, you are mentally moving around in this multidimensional space. You could even make a plot of properties of food along different sets of axes—for example, sweet versus salty, collapsing all the other properties down onto these two dimensions;

in making a plot of sweet versus salty, it doesn't matter how spicy or crunchy the particular food is.

Multidimensional mathematical objects work the same way—*generating* higher dimensional objects and manipulating them is straightforward. However, *visualizing* things in higher dimensions requires some mental gymnastics, and it wouldn't be wise to jump straightaway into doing backflips without warming up (or in my case, backflips wouldn't be wise to try under any circumstances). So, let's do a little warm-up exercise, with a simple circle. Your first challenge is to imagine (or draw) a circle and find a point as far away from the boundary of the circle as you can while staying equidistant from the boundary. Hopefully, you've landed in the center of the circle.

Here is the next step: go farther. It might take a moment to realize that to go any farther and stay equidistant from the boundary, you need to pop into the third dimension—into or out of the page. Visualizing popping into 3D isn't so hard for our 3D-wired brains.

But what about higher dimensions? We can do the same exercise as above, but now with a sphere instead. Imagine (or draw) a sphere and find a point as far away from the boundary of the sphere as possible while staying equidistant from the bounding shell, which will lead to a point in the center of the sphere. The next step is to go farther.

The cognitive experience I have when I do this last step feels like my brain knows what it wants to do, it just can't do it; there is almost a physical sensation of gears getting stuck. I would love to see modern fMRI imagery of people's brains trying to visualize higher dimensions. I will confess to exposing my kids to videos and imagery of higher dimensional objects at a young age, just in case.

In the meantime, we do have some other tricks to try to get our heads around what higher dimensional objects are like, which generally involves mapping them down to lower dimensions using projections, rotations, slices, and unfolding (and sometimes a combination of these). A necessary limitation with these techniques is that each of them loses information in the translation to lower dimension space, which we try to compensate for by rotating, projecting, and slicing in different ways.

To get a better intuitive sense for how these tricks work, I've found a good place to start is by flexing our 3D visualization abilities and envisioning how things would appear to an observer who was limited to 2D perception. Projections are a solid starting point because we regularly encounter them and the mental gymnastics are trivial. Here is an example of a 3D rectangular shell projected down to 2D: if you were only able to see the shadow of this shape, you would need to rotate the rectangle in three orthogonal projections to infer its full 3D dimensionality.

We can also take slices of an object. For example, if a 3D sphere were moving through a 2D surface, it would manifest as a circle that first grew larger and then shrank back before disappearing.

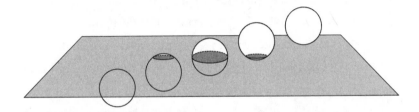

Finally, we can fold and unfold objects between dimensions. I absolutely hate folding things, especially clothes and sheets; if you share my dislike for folding, try thinking of this as a fun craft, like origami. If you lived in 2D, witnessing a cube fold and unfold in 2D would be baffling; it would be moving in and out of dimensions you are unable to perceive as though it were moving through itself.

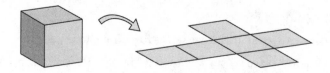

If we dial this up a notch, we can look at how a 4D cube would manifest in our familiar three dimensions. These hypercubes are also known as "tesseracts," which is a word and a concept I have held dear for many decades; tesseracts had a profound impact on my imagination as a child when I first read *A Wrinkle in Time*, in which tesseracts are a major plot device. I have often wondered whether reading this book had an impact on my interests and career trajectory—the power of children's literature should not be underestimated. To this day, I sometimes assign this book as fun extra credit. Go ahead, you know you want to read it.

We can watch rotating projections of 4D cubes in 3D space (and then mapped onto a 2D surface you are viewing), but sadly, printed media like this book are not (yet?) the best platform for video illustrations (YouTube will not let you down here, so a quick look for rotating hypercubes is worth your while if you have access to the internet at the moment). To my mind, projections of rotating higher dimensional objects look like they are turning themselves inside out.

We can also unfold this hypercube, as with the normal 3D cube. This unfolded hypercube is also known as a "Dali cross," after Salvador Dali's use of this form in his art (see his painting *Corpus Hypercubus*). Looking at this shape and trying to figure out how we might fold it into a 4D cube is perplexing, but our perspective on this is directly analogous to how a 2D being might perceive the folding and unfolding of a 3D cube.

While we are talking about hypercubes, we really must take a quick pause at one other waypoint—5D hypercubes. To spare both you and me, I am not even going to attempt to draw a 3D projection of a 5D cube. But . . . the two-dimensional projections of 5D cubes reveal some extraordinary behavior.

If you stare at the 2D projection of a 5D hypercube for a bit, you may see hints of 3D cubes appearing to poke in and out of the mosaic. This particular 2D tiling is an example of what are now called "Penrose tiles." As an undergraduate, the entrance of our math building was tiled with a Penrose mosaic—which I thought was brilliant (and now realize this is not at all unusual for math

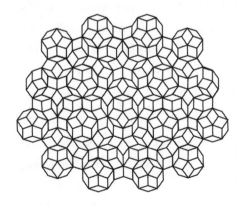

This mosaic is an example of a 5D hypercube crystal structure projected onto 2D for your viewing pleasure.

buildings, which doesn't make it less brilliant). If you tile a surface with only these two shapes in a set of Penrose tiles to infinity,
the resulting pattern can have rotational symmetry, and reflection
symmetry, but it will never have translation symmetry. While any
discrete pattern within the tiles could be repeated an infinite number of times, when an infinite plan is tiled, this tiling can never be
picked up, shifted in one direction, and perfectly line up. Thus,
this behavior challenges what we tend to think of as order—nice
tidy *periodic* patterns; while these tilings have order, they are *not
periodic*.

While Roger Penrose had this tiling named after him, examples of tilings like this were present in Islamic art back to at least
the fifteenth century in *girih* tiles. Given that the Middle East was
a hotbed of advanced mathematics at the time,[1] I find this temporal
coincidence to be intriguing. Some mathematicians have speculated
that these tilings were deliberately used in Islamic art to generate patterns that don't repeat, which would demonstrate an exceptionally
sophisticated understanding of these shapes that wasn't "rediscov-
ered" until the 1970s.[2]

Beyond merely having aesthetic value, Penrose tiles, like the set
above, have some extraordinary properties. For example, the number of possible tilings with these two simple shapes is *uncountably
infinite*, which I realize sounds redundant; in mathematical set theory there are both countable and uncountable infinities.[3] By analogy, we could imagine counting from 1 to 10 (heck, you don't have
to imagine), and you would have 10 numbers. But if we included all
the possible numbers between 1 and 10 including the nonintegers,
suddenly we have an uncountably infinite number of possibilities
as we expand the number of decimal points (things like 1.5872 and
6.2841). Related to their uncountably infinite number, the tilings

are also self-similar or fractal, which means that patterns can repeat within themselves at arbitrary small scales. As a side effect of these properties, if you lived in the 2D world of Penrose tiles and got lost, you could not determine where you were in the pattern, no matter how much you explored.

The shapes in Penrose tiles themselves are also fascinating; they have the golden ratio (phi or $\varphi = 1.61803\ldots$) baked into them in the ratios of their sides and center lines. We could spend an entire chapter discussing the number phi and why it is fascinating, but for the time being I'll just mention that, like its better-known sibling pi (π), phi (φ) is infinite and nonrepeating, and shows up in the natural world over and over through the ratios of lengths, the related angle, and the Fibonacci sequence. The number phi showing up in these curious tiles, which themselves are projections of 5D cubes, starts to feel like a set-up job.

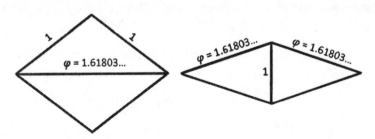

These tiles create what is termed a "quasicrystal," which is an object with order, but that lacks translational symmetry. A range of quasicrystals are now known to exist in the "real" world, which include coatings used for nonstick pans; if there is no periodic pattern of molecules coating the pan, it is much harder for food to stick. So, the next time you pull out a nonstick pan to make something delicious, you can think about how fascinating quasicrystals are, and how Penrose tiles are projections of 5D hypercubes.

Hints of Other Dimensions?

Given that we don't appear to be able to interact directly with other dimensions (if they even exist), you might justifiably be wondering what our motivation might be for considering them at all. To be clear, I would argue that all we really have to go on are some curious behaviors of things in the universe that might be easier to explain if we had extra dimensions laying around. This doesn't mean that these curious behaviors might not be explainable in some other way we haven't thought of yet.

I had a professor in graduate school who was fond of saying, "Be open-minded, but not so open-minded that your brains fall out," which never failed to induce gruesome imagery in my mind. His advice often percolates to the top of my thoughts when considering the possibility of other dimensions and where exactly in the brains-falling-out continuum he would say I am. However, I find that I am persuaded by the first of Arthur C. Clarke's "laws":

> When a distinguished but elderly scientist states that something is possible, [they are] almost certainly right. When [they state] that something is impossible, [they are] very probably wrong.[4]

On balance, I think I would rather err on the side of being open-minded than to be so calcified in my positions that I could not consider possibilities outside of the norm. I find I am most comfortable if I anchor myself in epistemology, and really try to weigh what we think we "know" alongside possibilities that we can't rule out. When it comes to other dimensions, I suspect my old professor would roll his eyes at me. As long as we are vigilant about what evidence we do—and don't—have and are equally mindful about what would be required to justify such a belief, I'm OK with pushing ahead and asking, "What if . . . ?"

With all of this in mind, here are some things about the universe that maybe, kind of, possibly provide hints of other dimensions. These range from imaginary numbers, to subatomic particle behavior, to more speculative ideas like string theory.

We've already encountered imaginary numbers in previous chapters—they have a way of showing up in these unsolved mysteries. The letter i shows up over and over again in fundamental physics—it is rampant in anything with oscillations—including electromagnetism and quantum mechanics. Recall that complex numbers, which have both a real and imaginary component, are actually *two*-dimensional. We *could* just think of this as a nifty mathematical trick that we happen to need (over and over again) to describe the universe, which is not an invalid perspective. Alternately, we might consider the position that if the natural world seems to require an extra orthogonal dimension to describe it, maybe there is an extra orthogonal dimension. To be clear, although I think this idea is alluring, I am fairly agnostic about whether imaginary numbers are just a nifty trick or reflect something deeper—regardless, our standard "real" numbers don't seem to be up to the task of explaining reality on their own. If imaginary numbers *do* reflect something deeper and we don't give this concept due consideration just because the idea sounds fanciful, then we are doing ourselves a disservice.

One example involving complex numbers that I find intriguing is particle spin. Explaining this requires a little preparation, but we will end with Möbius strips looping into other dimensions, so I think it is worth it.

Here is the pared-down physics that we need to get to the cool part: elementary particles have intrinsic angular momentum that we call "spin." The concept of "spin" can be misleading in this case because particle spin is not the same as an actual object spinning in normal life, like a top, a coin, or a ballerina; since the elementary particles are point-like, rotation in space doesn't really make sense. Rather, in this context, "spin" indicates that the state of the particle oscillates through phases. Given the discussion above, you may have guessed that these phases turn out to be complex values.

There's more, and it gets weirder. Not only do these point-like particles have angular momentum, but the allowed values of angular momentum are set in stone. In the normal world, you might watch a spinning top, coin, or ballerina, and the spin could change speed (along with the angular momentum). For elementary particles, the spin *must* have a specific predetermined value—it can neither slow down nor speed up. Ever. These spins have a quantum number associated with them that is always n/2, where *n* can be any nonnegative integer (in principle). There are particles out there in the wild with spins of 0, 1/2, 1, and 3/2 (and possibly higher, but I don't want to take us down that path right now).

What exactly do these spin quantum numbers mean? Now the situation gets extra weird. The *inverted* spin quantum number tells us how much rotation the particle has to go through for its phase to line up again. A particle with a spin 0 is entirely symmetric—it doesn't

Particle spin	Analogy	Rotation needed to realign
0	⊙	Any
1	⊕	360°
2	❶	180°

matter how its phase is oriented. A particle with a spin of 1 needs to rotate once around (360 degrees) to have its orientation return to the starting position. So far, so good.

Then we get to particles with spin quantum numbers of 1/2, which include things like electrons and protons, so we can't just ignore these and pretend they are odd ducks. If a particle has a spin quantum number of 1/2, that means it must rotate around *twice* before the phases line up again.

Particle spin	Starting position	1 rotation	2 rotations	Rotation needed to realign
1/2	⇧	⬆	⇧	720°

For "normal" things like the top, coin, or ballerina, this makes no sense at all. If you stand up and spin around one full rotation, you will return to where you started. But this is just not how these spin 1/2 particles work. The best analogy I have is that of a Möbius strip. Möbius strips have all sorts of fun mathematical properties, including the fact that they only have one side, and they are nonorientable—which is a fancy math term meaning that you can't distinguish between clockwise and counterclockwise. If a 2D person lived in a Möbius strip, after one rotation around, they would find themselves upside down.

To accomplish this, the Möbius strip has to twist around in the third dimension—this is equivalent to the work of the complex values in phases of particles. Whether or not the particle spin taps into a higher dimensional space, mathematically *they behave as if they do.*

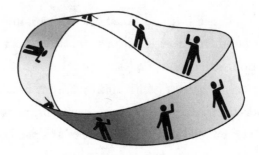

In case you are curious, and I hope that you are, there are also 3D versions of Möbius strips called "Klein bottles," which twist around in the fourth dimension. Like many higher dimensional objects, when we try to visualize these in mere 3D space, Klein bottles appear to intersect themselves in 3D. As with Möbius strips, Klein bottles only have a single side and don't enclose any volume—the "inside" becomes the "outside" in the same way that one "side" of a Möbius strip becomes the other, which is a pretty nifty trick.

Our next hint at other dimensions comes from the fact that general relativity and quantum mechanics don't get along, which has been a thorn in our sides for many decades. As we encountered in the chapters on the Big Bang and black holes—our physical understanding in both of these extreme scenarios hits a wall because we don't yet have a theory of quantum gravity *that can be tested*. Never fear—we may not (yet) be able to test our hypotheses, but we do have them. One of the leading contenders is a theoretical framework called "string theory." (FYI: This is a prime example of the word "theory" being misused; despite being right there in the name, "string theory" doesn't yet qualify as a "theory" in the formal scientific sense because it hasn't been tested—at all. However, I will concede that "string hypothesis" sounds a tad inane.) The central idea in string theory is that fundamental particles are really vibrating strings of energy.

Exactly how the string vibrates determines the properties of the particle at hand. If you've ever played a stringed instrument—or even seen one being played—you are familiar with the concept of vibrating strings, and the exact form of their vibration can change the tone from being dulcet to shrill. If you've ever been to a children's orchestra concert, you know exactly what I am talking about. As the parent of a cellist and a once-upon-a-time violin player myself, I have been on both sides of this. When a player changes the length of the string that is allowed to vibrate by putting their finger down, the pitch of the note changes. Depending on the length of the string, only certain wavelengths of vibrations are allowed because of what we call "boundary conditions"—the string is physically affixed at two ends, so an integer number of wavelengths must fit between these ends.

The same principle applies for strings that are in a loop—by virtue of being in a loop, the string must connect seamlessly back to itself, so an integer number of wavelengths must fit around the loop. The "strings" of string theory are not so different—except for being unfathomably small, and (perhaps much to the envy of stringed-instrument players everywhere) these strings are thought to vibrate *in other dimensions*. In the currently favored flavor of string theory, called "M-theory," there are a total of eleven dimensions in our reality—the familiar 3 + 1 of our everyday space and time, plus an additional seven. My favorite number happens to be seven, so I will take is as a personal victory if this number turns out to be correct (kind of like taking pride in your favorite sports team winning, despite the extent of your participation taking place from a couch). However, if you have an ounce of skepticism (and I hope you do), you might be wondering where the heck these other seven dimensions are; presumably you have gone through your life without noticing extra dimensions lying around.

In the case of string theory, the idea is that six of these extra dimensions are hidden from both our own human senses as well as the ability of our lab equipment because they are "compactified" (which spell-check insists on telling me isn't a word—I promise this is the legitimate term we use). The main issue I have with the word "compactified" (aside from spell-check annoying me) is that it might suggest that something (in this case the strings) was actively made to be compact by some outside force—for example, "I compactified my clothes to fit them in my suitcase." Or "I compactified the loaf of bread by sitting on it" (which, for the record, a person in my life, who will remain anonymous, loved to do as a child). In the case of strings, all we mean by the word "compactified" is that they are *really* small. Like almost at the Planck scale. So small that we cannot perceive or probe them with current technology.

When I am teaching this material, at this point I will offer my students the option of doing an extra-credit assignment, which I've found to be a good way to play around with how something with compactified dimensions might work. To be clear, what I'm about to suggest isn't truly compactified—it will still exist in our 3D space (if your process turns out otherwise, probably you should write that up for a physics journal), but by analogy I find that this activity can help with intuition. Also, I just like doing art, and giving my students an excuse to flex their creative powers amid intense existential discussions is a welcome break. What I want to encourage you to make are "hexa-hexa-flexagons," for which there are numerous patterns with instructions online. I have my own favorite hexahexaflexagon, which I spent the better part of a day designing and coloring many years ago; it is now so well loved that the folds are starting to fray. One day my husband was absentmindedly fidgeting with it and somehow managed to get it into a hybrid state I didn't know was possible. So, kudos to him.

Hexahexaflexagons behave like compactified dimensions in the following way: to an external observer (especially from far away), they look like 2D objects, with a front and a back. However, as you fold

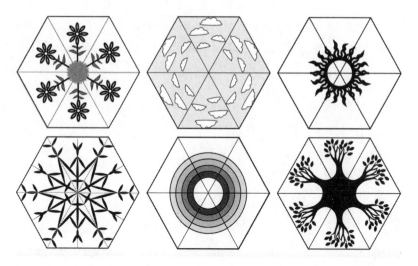

An example of the six faces of a hexahexaflexagon that are "compactified"—only two of which are apparent to the outside world at any given time.

and invert them, *four other sides* become apparent. In other words, these apparently 2D objects, have six "sides." This analogy isn't perfect, and I am sure there are folks who do serious topology work in math ready to argue about how many sides hexahexaflexagons actually have, but the analogy works well for illustrating how dimensions might exist in an invisible compactified way.

A leading mathematical framework for the form these extra six compactified dimensions might take is known as a "Calabi-Yau manifold." ("Manifold" means that any little part of it can behave as if it were flat. But you don't need to worry about that now—unless you are a math major.) These manifolds have properties that are very compelling, including important symmetries and being consistent with Einstein's field equations (formally, they are "Ricci flat"). I'm sure you will be shocked to hear that they are described by complex numbers. Curiously, these manifolds have three spatial dimensions, which may seem familiar from your everyday life. However, since these three spatial dimensions are *complex*, each of them is really two-dimensional, so we end up with a total of six extra dimensions. The shapes that result are stunning (at least when projected into 3D space and then onto 2D paper).

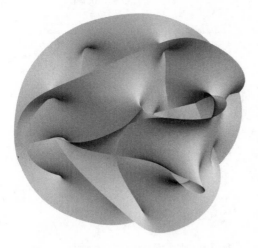

A 2D projection of a Calabi-Yau manifold. Credit: Image modified to black and white from https: //en.wikipedia.org/wiki/File: Calabi-Yau.png, in turn modified from A. Hanson, "A Construction for Computer Visualization of Certain Complex Curves," *Computer Science*, 1994.

You may have noticed that "shapes" in the above sentence is plural. There are possibly an infinite number of possible Calabi-Yau manifolds (which is unresolved at the time of this writing), each of which results in different particle properties. If the properties of the particles arise from the exact shapes of these compactified extra dimensions, then these properties are literally stamped into the very fabric of space-time. The catch is figuring out which of the possibly infinite number of Calabi-Yau manifolds might correspond to our universe. If we do figure out which manifold yields the properties we observe, the next obvious question is *why that particular manifold* out of all the possibilities? And thus, again we end up with a stack of turtles.

One hypothesis that theoreticians are playing around with is that before cosmic inflation (see the chapter on the Big Bang), all the dimensions of the universe were compact and roughly on equal footing. When symmetry was broken and inflation took off, the dimensions now envisioned to be compact simply did not inflate (exactly why they did not inflate is a necessarily open question since we don't even know that they exist at all). This also begs the question of whether other possible universes could have inflated with different dimensional compactifications.

The fact that we can't (yet) probe these ridiculously tiny size scales also means that we can't *directly* test this theory, so your skeptical radar should be on high alert. As physicist and mathematician Peter Woit argues, string theory is "not even wrong" in allusion to a famous quote from Nobel laureate Wolfgang Pauli.[5] To an outsider, the phrase "not even wrong" may seem innocuous, but to Pauli this was a grave insult, indicating that a hypothesis was not even in the realm of science, which requires the ability to falsify.

That being said, there are a number of indirect ways we might gather *circumstantial* evidence for string theory; however, none of these would be "proof" in themselves, and moreover, after decades of trying, even these indirect methods have not yet borne fruit. For example, when the Large Hadron Collider smashes particles together at unfathomably high energies, it is possible that micro–black holes could form—but only if there are compactified extra dimensions that would focus the gravity. In our merely 4D universe, a collider would need to reach energies of about 10^{28} eV to create micro–black holes, while the current collider limits us to energies below about 10^{13} eV; so, we are roughly fifteen orders of magnitude away from creating micro–black holes *without* extra dimensions. However, some predictions indicate that with extra compactified dimensions, the creation of micro–black holes might just be within reach of current collider technology.[6] Rest assured that if the Large Hadron Collider had managed to create micro–black holes, you would have heard about it, probably even above the fold in your favorite newspaper.

Despite the challenges with string theory, a legion of ordinarily skeptical and proof-demanding mathematicians and physicists have an almost religious belief that string theory (or something very similar to it) *must* be true. What gives? In one fell swoop, string theory could unify the fundamental forces, provide a theory of quantum gravity, and explain why particles have the properties they do (which is important in the chapter on fine-tuning). That is already a lot of heavy lifting for a single theory, but in case that weren't enough, string theory possesses a coveted quality in mathematics—beauty. It is OK to admit if you are having trouble reconciling the words "mathematics" and "beauty" sharing a sentence. When we talk about things being "beautiful" in math, we are referring to the underlying concepts, not the contrivances we humans have developed to try to manipulate these concepts in ways we can understand and write down on paper.

To be fair, the beauty of a mathematical concept is surely subjective and dependent on what we humans tend to find aesthetically pleasing, which in turn is ultimately rooted in evolution (which has a lot more to do with procreation and survival than ultimate truths in the universe). In math and science, beauty is generally determined by symmetry, simplicity, and elegance. These attributes suggest an underlying organization, coherence, and harmony that we often infer to indicate a deeper truth, which is how we end up with so many mathematicians and physicists saying things along the lines of "Beauty is a guide to truth." As we will encounter in the chapter on laws, there is a broad implicit belief that the underlying architecture of the cosmos must be elegant.

Many scholars (including myself) believe that we *discover* underlying concepts as opposed to *inventing* them. If one holds the belief that the underlying truths of the universe must be beautiful, when you come across a concept like string theory, which is ripe with symmetries and elegance, it feels like it must be true (and if it isn't, it should be). The bottom line is that string theory would not only solve some of the major outstanding issues in physics, but it would do so with beauty. In terms of the justified-true-believed framework of epistemology that we saw in Chapter 2, one can argue about whether "beauty" should count as a form of justification. Most scientists I know see beauty as suggestive, but not hard evidence. The upshot is that string theory still sits in the realm of "not even wrong," but that may well be a lesser insult than saying it is "not even beautiful."

You may have noticed a problem with my math in the last section: if the familiar universe we experience has 3 + 1 dimensions, and Calabi-Yau manifolds have six compactified dimensions, why does M-theory have a total of eleven dimensions? The answer to this question leads us to mathematical constructs called "branes." You can think of the word "brane" as being short for "membrane," but these geometrical structures could have any number of dimensions. M-theory invokes branes to provide strings a place to hang out and have fun (and also unifies six different versions of string theory), but this comes with one extra bonus dimension (in addition to the 3 + 1 + 6), ultimately suggesting a total of eleven dimensions.[7]

In the brane cosmology scenario, you can imagine the three dimensions of space we are familiar with as a brane in a higher dimensional space, meaning this extra dimension that gets us to eleven is *macroscopic*—not one of these tiny little curled-up manifolds. A very cool feature that naturally falls out of the brane cosmology conjecture is that it could provide a reason for gravity to be so ridiculously weak. Specifically, in these models, the type of string responsible for gravity is not confined to the brane in the way the other types of strings are. If gravity can leak off into other dimensions, we have a handy mechanism to explain why this force is so singularly weak.

There is a tiny problem with these models, though—a couple decades ago the models had the benefit of not being testable, but today they are no longer "not even wrong." Gravitational wave detectors are now up and running—and you will be glad to hear they are, in fact, now routinely detecting gravitational waves. Occasionally, we can also detect the source of the gravitational waves with good old-fashioned photons, in which case we can do all kinds of helpful tests on how gravity behaves (because we can "see" what caused the waves to begin with). With the data currently in hand, limits are starting to be placed on whether gravity could be leaking into other dimensions, and the results are not encouraging for proponents of these brane models.[8] That being said, as we get more data the situation may become more clear.

Even if these particular brane models don't turn out to reflect reality, they have us thinking outside the box (or outside the brane in this case). This is a great illustration of my favorite way of doing science—sometimes seemingly crazy ideas turn out to be true. And even if they are not true, we stand to gain a lot of insight along the way. Dead ends in science happen all the time—we scientists often joke about creating a journal for "failed" experiments (Lord knows I would have a lot of publications there). Since we generally don't publish our dead ends, and because these dead ends almost never make the popular press, it can appear to the outside world that science is much more linear than it actually is. Instead, we often backtrack, go in loops, and rekindle old ideas that were once thought incorrect. If we only did experiments that we knew would work, it would be very hard to make progress. In fact, one of the mottos I lean into frequently is "Some experiments fail," which I invoke not only with my own research, but also cooking, painting, course development, talks, and social interactions.[9] I maintain that if all your experiments succeed, you are not thinking creatively enough or trying difficult enough things.

Not only might there be other dimensions within our universe, there may even be other universes—with their very own dimensions. The possibility of a multiverse takes us straight to the borderlands of science and the intersection between science and philosophy. To be clear, there are scientists who will firmly argue that this topic is solidly encamped outside the realm of science, and I am somewhat sympathetic to this position—once again for the people in the back, something must be *testable* to be able to carry a scientific passport. However, I also maintain that not considering the possibility of other universes—just because the idea might not be currently testable—is myopic. If we don't even let ourselves think about eccentric ideas to begin with, our apparent inability to test them becomes entirely self-fulfilling. Unless one wants to argue that our current physical understanding, mathematical frameworks, and technological capabilities are the pinnacle of what is possible, I think we need to be mindful of our stance on what may or may not be possible in the future.

Entire taxonomies have been developed around different instantiations of other universes, some of which are fantastical and rely on the concept of other dimensions, others of which seem downright prosaic to me. Many of the differences between the multiverse concepts come down to what we consider to be a "separate" universe. For example, when talking about "another" universe, do we simply mean there is no causal connection with our universe? Or do we mean this other universe exists outside of our familiar 3 + 1 dimensions? This brings us to yet another semantics issue, which seem to always come up with topics that are not naturally part of our normal human experience; how are the concepts of "universe," "multiverse," and "metaverse" related? My experience is that these terms are not used consistently out in the wild, and now the term "metaverse" is being co-opted by virtual reality, which complicates things even further. For the sake of starting somewhere and trying to at least stay internally consistent here, I am going to adopt some definitions by fiat (which I have no ordained authority to do). So, if you come across these terms in other contexts, be aware that they may be used differently.

> *Universe*: A topologically connected area of specific
> dimensionality. For example, in our familiar 3 + 1 universe,
> there may be regions outside of our cosmic horizon with
> which we share the same underlying dimensional framework
> of space and time, even if we cannot have causal contact.
> Nevertheless, the topology of space-time is smoothly and
> continuously connected between regions.

> *Cosmos*: I will use this to describe everything that is, whether
> we can observe it or not, and regardless of whether it is
> causally or dimensionally connected to our universe. As far as
> I'm concerned, this is the top of the hierarchy.

> *Multiverse*: Distinct regions of the cosmos that are either
> topologically disconnected and/or otherwise physically
> distinct.

> *Metaverse*: A parent universe from which other universes
> might be generated.

Here are some toy multiverse ideas to consider. To my mind, the most mundane scenario simply acknowledges regions beyond our cosmic horizon: in the working definitions above, these don't really even qualify as other universes. The idea here is based in the reality that we are only causally connected to a limited region of the universe, which we refer to as the "Hubble volume." You might be inclined to think that as time passes, our Hubble volume would increase—after all, if there has been more time for light to travel, we ought to be able to detect light from farther away. This is true, but there is another effect that trumps this; because the universe is expanding (and indeed accelerating, which we will come to in the next chapter), the farther away something is from us, the faster it is receding from us. At a specific distance from us, things in the universe are moving away from us at the speed of light; this distance is the radius of the Hubble volume (currently a bit more than 14 billion light-years), and beyond this radius we cannot ever have causal contact (unless someone comes up with a method for superluminal travel). As time goes by, more and more of the universe will move outside of this Hubble volume because the universe is expanding at an accelerating rate. If you follow this scenario to its logical end, the future is quite grim—eventually we will be completely alone in our Hubble volume. Of course, whether we might survive that far into the future is an open question. I sometimes imagine an astronomer of this distant future and wonder what they might think of the universe if there were nothing but us in our own little provincial Hubble volume to observe.

If you want to ignore my working definitions above, we could consider anything outside of our own Hubble volume to be "another" universe (and some practitioners do). Regardless of the semantics, we do not know (and can never know)[10] what is going on outside of our bubble. This physical interpretation of regions outside our Hubble volume not being causally connected to us is generally uncontested in the field. Since we can't ever observe outside of our Hubble volume,

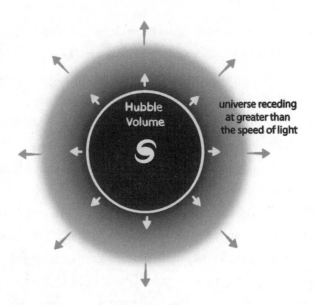

we must concede the possibility that the cosmos might behave very differently out there (which is crucial in the next "multiverse" model), and we might be in some anomalous location. If this were an old-fashioned map, "There Be Dragons" would be written (no doubt in lovely calligraphy) on the outside of the Hubble-volume circle.

There is a stronger version of this "multiverse" theory, though, which spices things up a bit. If the universe is infinite, then it must contain an infinite number of Hubble volumes. If this is the case, then every possible set of initial conditions will not only be manifested, but it will be manifested an infinite number of times. This scenario gives rise to some obvious and disconcerting ramifications—in particular, this would mean that there are exact copies of our Hubble volume out there in the cosmos—an infinite number of them. The implication being that there are an infinite number of you, me, and your first-grade teacher. In fact, we don't even need the universe to be infinite to expect identical copies of our Hubble volume to be out there; we just need the universe to be sufficiently large—large enough to require repeating of exact initial conditions. According to physicist Max Tegmark, we might expect the nearest identical copy of our Hubble volume to be $(10^{10})^{115}$ meters away.[11] To be fair, that is an unfathomably big number, but it is also not infinite.

A close sibling to simply having other Hubble volumes is the chaotic inflation hypothesis: this idea invokes the possibility of distinct physics, even if the underlying dimensionality is topologically connected. In chaotic inflation (also referred to as "eternal inflation") not only are there other Hubble volumes, but due to quantum fluctuations in the early universe, different patches would have inflated differently. Moreover, regions of the underlying "meta" universe will continue to spawn new bubble regions with essentially fractal behavior, with each bubble's local inhabitants justified as thinking of their own bubble as the "universe."[12] In this scenario, different physics might also manifest in each bubble;[13] recall from discussing string theory that in this conjecture exactly how compactified dimensions do (or don't) inflate and the forms they take directly impacts forces and particles properties we observe in the universe. As a result, chaotic inflation has direct implications for the fine-tuning problem we will discuss soon.

Chaotic inflation also leads to what physicist Alan Guth calls the "youngness paradox," which is one instantiation of what is termed the "measurement problem" in eternal inflation—when we are dealing with infinities, exactly how we decide to make measurements can radically change the outcomes. In this case, in a metaverse that is constantly inflating new pocket "universes" at an exponential rate, the number of newly formed universes will vastly outnumber the number of more mature universes. To imagine this, let's think about rabbits. A typical rabbit can start reproducing at about six months of age. If we start with just two rabbits, and the average litter has six offspring, within two years we will have about eighty rabbits—and the average age of all these rabbits will be only a bit more than two months. In exponential growth like this, the "babies" will always radically outnumber the "adults," which means that if you were to select a rabbit (or a universe) at random, it is most likely to have just been born.

The so-called "paradox" arises if we want to assume that we are typical observers—the principle of mediocrity yet again. Yet here we are in a universe that seems to be not quite 14 billion years old, when the "typical" universe in chaotic inflation would have just been created. This quickly takes us down the rabbit hole of the anthropic principle—that the properties of the universe must allow for our existence, which then leads to the question of whether life could have arisen in any form in a much younger universe. Taken ad nauseam, we land on the concept of Boltzmann brains (speculative self-aware entities that have a finite probability of occurring from random fluctuations—even in the very early universe), which would radically outnumber sentient life similar to ours in eternal inflation scenarios.

It turns out that it is very hard to create models of cosmic inflation that *do not* result in eternal inflation and a multiverse of this type. Alan Guth is even on record as saying, "It's hard to build models of inflation that don't lead to a multiverse," so the "youngness paradox" is not something we can simply sweep under the rug.[14] There are, of course, solutions—one of the most obvious being that our pocket of the universe is not typical, which in turn opens a whole big can of worms. Another option is that we still don't understand the physics of inflation, which isn't terribly surprising, given that inflation convolves a lot of physics that we know we don't understand. In a paper submitted by the late Stephen Hawking (shortly before his death) and coauthor Thomas Hertog, the duo put forward a revised inflation scenario that simplifies the situation, but it is still not a panacea.[15] According to Hawking, "We are not down to a single, unique universe, but our findings imply a significant reduction of the multiverse, to a much smaller range of possible universes."[16] So, I guess that is progress.

A second cousin to chaotic inflation is the idea of fecund universes. In the chapter on black holes, we saw that one of the possibilities for what happens inside them is the creation of a whole new "universe." The upshot of that section was that we could very well be living inside of a black hole created in a parent universe. To be sure, this idea is not without criticism, some of which I am sympathetic to and other bits of which I am not.[17] For opponents to the black-hole-universe idea, the current crux of the matter hinges on the black hole information paradox and whether black holes irretrievably swallow information. Without any information, we may not have the raw material we need to make a new universe, which of course begs the question of where the information that seeded our universe came from to begin with.

If any version of the fecund universe hypothesis plays out, it is possible that every black hole created in our universe generates the birth of a new "universe." If the throat of the wormhole connecting them pinches off (very much like an umbilical cord), then parent and child universe could become topologically separate. In the process of this child universe being created, the properties of particles and forces in the parent universe might be passed down and/or mutated. For example, you could imagine different variations of a Calabi-Yau manifold being imprinted into the dimensions of the baby universe. In this case, there could be entire genetic lines of universes, with some branches effectively sterile and others reproducing with gusto.

We can also conceive of other universes that are not, and never have been, topologically connected to our own, which leads to brane-world cosmology. This version of a multiverse connects directly back to string theory and M-theory.[18] But more than a decade before M-theory was around, scientists were thinking about a potential cosmology in which our universe might be on a "domain wall" in higher dimensional space.[19] As a reminder, the underlying idea in braneworld cosmology is that our universe is equivalent to a multi-dimensional membrane that exists in a higher dimensional space (which we refer to as the "bulk").

Braneworld cosmologies do a lot of nifty things, including help explain why gravity is weak, and they even give rise to possible explanations for the Big Bang, which you might recall from that chapter (in particular, the "ekpyrotic scenario" gets into what might happen if branes collided in this hyperdimensional space).[20]

Unfortunately, experiments done to date are putting constraints on possibilities of braneworld cosmology, including results from both the Large Hadron Collider and gravitational wave detectors.[21] I say "unfortunately" because this is one of those hypotheses that I would really like to be true, but for better or worse, the universe doesn't always do what I want. As is always the case with confirmation bias, the more we want something to be true, the more skeptical we should try to be; in this case I have to actively force myself not to exploit the loopholes for this conjecture to stay viable. But who knows? I'm not totally giving up hope yet.

Quantum mechanics, much like string theory, also offers a way into new universes through the many-worlds interpretation. Quantum mechanics tells us that things exist in a quantum mechanical superposition of states (they both exist in every state and also in no state simultaneously, with appropriate probabilities applied to every option). These quantum things don't make up their minds about their state until they are forced to by interacting with the macroscopic world. To be fair, sometimes I feel this way going into an ice cream shop; in advance I could assign a probability that I might pick any given flavor, but it isn't until I'm standing there looking over the options that I am forced to decide. In the Copenhagen interpretation of wave function collapse, I do truly have to decide, even if my choice is entirely random.

The thing I love about the many-worlds interpretation is that (in this analogy) I get to sample every single flavor of ice cream, albeit in a different universe that branches off and of which I have no awareness. Instead of wave functions collapsing to a random single value with no rhyme or reason, in the many-worlds interpretation a wave function undergoes quantum decoherence to every possible value, and in the process spawns off a new universe with that property. A minor issue that many people have with this scenario is that we seem to end up with an infinite number of universes unless we can find ways to limit the branching or otherwise merge paths.

One aspect of this hypothesis that I find intriguing is that the resulting range of universes are effectively similar to the beyond-the-cosmic-horizon scenario if the universe is infinite. In both cases, every possible outcome from every possible quantum configuration and wave function collapse becomes a reality that exists outside of our ability to interact with it.

To be sure, the many-worlds interpretation leaves open a range of existential questions, about which I am as clueless as anyone else. For example, as these new universes branch off, why do we find our consciousness in this particular one, and what does that mean about consciousness?

This is a nonexhaustive list of ideas for other possible universes (and I hope you are not exhausted), but it touches on some of the main ideas that people are seriously considering, with things like actual peer-reviewed papers. An essential point to remind ourselves of is that something does not necessarily need to be observable in order to make predictions that are testable. This may seem a bit oxy-moronic, but this principle is engrained in how science proceeds. Gathering circumstantial evidence from predicted properties and behaviors that *are* observable is also pretty much the only option we have in astronomy—even for topics as pedestrian as stellar evolution; we can neither watch stars evolve over human lifetimes, nor can we send probes to their cores to see what is really happening. Instead, we develop models based on the physics we understand and make predictions that can (hopefully) be testable.

One question we must continually ask is whether there are other models that could produce similar predictions, in which case we must keep an open mind about which model (if any) might be correct until we come up with a test that can discriminate between them. A fundamental ingredient of these models is that they are rooted in our understanding of physics—creating models for situations that extend beyond our current understanding of physics is where we start to get queasy. Creating models that extend beyond our current understanding of physics and which do not make testable predictions will make even the most credulous scientist turn tail.

In the early years of teaching this course material, I would ask students to put forward concepts on how the crazy ideas (like multiple universes) might be testable as part of their standard homework on the weekly topic. My hope was that by overtly inviting students (who had not yet been fully baked into cynics) to consider possibilities, they might be empowered to generate unconventional ideas for approaching apparently intractable issues. In hindsight, my aspiration was overly naive, and the students generally did not have enough scaffolding to get traction on topics like multiple universes.

To be sure, there are a lot of crank ideas out there, but dismissing the entire area of thought because it is rife with crazy ideas doesn't mean that there isn't some genuine truth and understanding to be gained from careful consideration. As long as we are mindful of what is or isn't testable (or one day might be), I think we are on solid enough ground. One of the paths forward *scientifically* is to ask ourselves how an idea might be testable—even if that technology doesn't currently exist.

I'm afraid we are left in in a pretty unsatisfying state with the question of whether there are other dimensions. We don't know that there aren't other dimensions or a multiverse, and we have a handful of hints that there might be. The existence of other dimensions or universes is also tied to the next two chapters—why our universe has the laws that it does, and whether the parameters of our universe are fine-tuned. The real trick is in trying to figure out if or how we could interact with these other dimensions. The ideas we have thought of so far have come back empty-handed. A positivist might argue that if we, indeed, cannot interact at all with these other dimensions, whether they exist doesn't matter. In terms of the purely practical, perhaps that is true. But I don't think I'm alone in having a more ontologically oriented mindset—I want to know what *is*, not just what we can perceive.

Interstition

We are working our way deeper into the landscape in which the great mysteries of the universe dwell. We appear to live in a universe that allows for black holes to exist and something like a Big Bang to have occurred. We have a singular dimension of time that seems to enable anything to happen at all. At present day, it appears that we have four distinct forces in the universe, without which life-as-we-know-it would be impossible.

But why does our universe have any of these characteristics to begin with? Why do we have the forces that we do and the fabric of space-time in which these forces operate?

The question of why we have the laws of nature that we do underlies all of physics—and indeed our very existence. The next two chapters delve into what the laws are, where they came from, whether they might have been otherwise, and whether our universe would be habitable if they were.

Because we are embedded in the universe made from these laws, we are faced with a paradox of self-reference, and the path out of this morass is not clear. At least not yet.

9

What Determines
the Laws of Nature?

IT WOULD APPEAR THAT OUR UNIVERSE HAS SOME KIND OF underlying laws or rules that govern what happens. If this were not the case, there would be no rhyme or reason to anything, no order, and no predictable behavior. Such a lack of underlying rules would manifest in myriad ways that seem unfriendly to life, to put it mildly; for example, if there were no law telling things how to respond to gravity, there would be no stars, no orbits for planets even if there were stars, and no planets to begin with (more on this in the chapter on fine-tuning).

Given that we are embedded in these laws, and indeed a product of these laws, it is extraordinarily easy to take these laws for granted. We are readily lulled into a sense that these laws *just are* and give them no further thought. To me, just accepting these laws as God-given feels a lot like a grown-up responding, "Because that's the way it is," to a child's question. I hope you are as deeply unsatisfied with that answer as I am.

The laws of nature are a fundamental aspect of our ontological landscape, but we have no idea why we have the laws that we do, or whether they could have been otherwise. The universe clearly seems to have rules, but we don't know their source, where they are encoded, or how they inform the universe about what to do. We interpret the laws of nature as they manifest in forces, but the relative strengths of the forces seem totally random, lacking any underlying symmetry that we physicists have come to expect. There is a deep belief among physicists that all four known forces should be unified, but it isn't clear how. Nothing like a good ontological crisis.

Thinking about something as abstract as the *laws of nature* having an existence will also have us chasing our tails and getting stuck in a quagmire of infinite regression (as usual): Are there metalaws that govern what laws our universe can have? In which case, what determines those metalaws?

Trying to understand the universe we ourselves are living in is nontrivial. I marvel at the physicists who spend decades honing highly sensitive (and typically commensurately expensive) experiments, just to get yet another negative result. I want to pause here and take a step back to get a little perspective on what we are trying to do. When I teach about this in a classroom, I have the luxury of a blackboard I can use in real time, a projector to show animations, and also students who are more or less obligated to bear with me as this lesson unfolds. Trying to cover this in a static book may be a fool's errand, but to my mind the end goal is meaningful enough to warrant the attempt.

This may be an odd request from an author to a reader, but I ask that you try to trust me for a few pages and know that this is going somewhere (but I don't want to tell you where just yet, because I think it is more impactful if you go on the journey to get there). OK?

This excursion requires us to enter the world of the Game of Life. Not to be confused with the board game that has you making major life choices with the spin of a wheel, the Game of Life we consider here was developed by mathematician John Conway in 1970 (and therefore often called "Conway's Game of Life").[1] Explaining this game will take a hot second, and reading about it frankly doesn't do it justice—but the good news is that there are several Game of Life simulators available online that you can (and should) test out and play with as we go on this side quest. My students—who are way more video game savvy than I am—confirm the game is unexpectedly compelling.

The brilliance of the Game of Life is its simplicity. It takes place on a basic grid of pixels (or cells), like you might find on a standard piece of graph paper (which was a go-to in the 1970s when home computers were far from common), and the cells in the grid can either be "alive" (filled in) or not alive (empty), with each cell having eight neighbors (diagonals count). Moreover, there are only two simple rules that govern what happens:

(1) Cell reproduction: the birds and bees of cell birth requires not two, but three living cells—if three living cells are touching an empty cell (and corners count as touching), the empty cell will become alive in the next time step.

(2) Cell death: cells can perish from either overcrowding or loneliness—if a living cell has fewer than two living neighbors (remember that corners count) or more than three living neighbors, it will meet its demise in the next time step.

That's it. I absolutely realize that playing this game may sound like watching grass grow, and I will be the first to admit that sometimes mathematicians find curious things amusing, but *something* made this game take off with a cultlike following, and countless players spend untold hours experimenting with this game, and it wasn't the cutting-edge graphics. Of course, this was also the era in which *Pong* was all the rage, so that tells you something about the quality of graphics that impressed us.

To get warmed up, let's try a few straightforward examples. Consider the grid below with three distinct patterns in it (A, B, and C). Given the two rules above for birth and death of cells, can you work out what will happen to each pattern in the next time step?

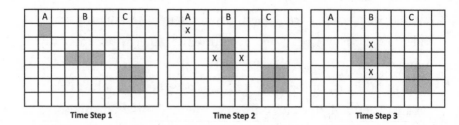

Time Step 1   Time Step 2   Time Step 3

Pattern A is straightforward; it has no living neighbors, so it will perish from loneliness. Pattern B is slightly more complicated, we must consider each of the three cells in turn—as well as the surrounding cells. The cells on the left and right of Pattern B will die of loneliness (they each only have one friend), but the cell in the middle just keeps living its life like nothing has happened. On the other hand, the central cells directly above the middle of pattern B each have three living neighbors, so they will come to life in the next time step. Pattern C is quite stodgy—each of the four cells in this pattern has three living neighbors, so this pattern is just going to sit there unchanged in the next iteration.

For completeness, let's do one more time step. Pattern A has gone extinct, so we can mourn it, but we must go on with our lives. Pattern B will keep oscillating back and forth between being a horizontal and vertical line of three cells (this pattern is called a "blinker"), and Pattern C will just sit there in perpetuity. (For obvious reasons, this pattern is called a "block." With such evocative pattern names, clearly mathematicians are as adept at naming things as astronomers.)

That is all well and good, but these examples don't really explain why this game is so compelling. There really are people who spend extraordinary amounts of time playing around with the Game of Life. The key to this game is the initial conditions—exactly how you populate the board to set it up can have astonishing outcomes.

When I have the luxury of a projection screen in class and not just ink in a book, I will do a few iterations of the following, and I am keen for you to do this on your own if you can:[2] just randomly scribble on the grid and then press "Play." Most of the time what unfolds is initially a bunch of chaos that quickly settles down into a few key components, almost always eventually leaving the grid with only patterns that stably oscillate (like the blinker) or patterns that just sit there and do nothing (like the block) and a handful of cute little patterns called "gliders" that escape off into the grid universe somewhere. Watching these patterns unfold is unreasonably entertaining (at least to me, although I realize that my level of nerdiness may call my judgment on such issues into question, but hundreds of appropriately skeptical and definitely not nerdy college students agree with me). Really, you just have to try it for yourself.

This is where the initial conditions kick in. Even when populating the grid randomly, something truly interesting can happen—perhaps some type of extremely complex and stable pattern. You can also populate the initial conditions with specific intention to generate patterns that do extraordinary things. For example, there are patterns that can grow indefinitely, patterns that generate "stargates" within the game (transporting spaceships faster than physically allowed by travel in the grid), patterns that generate prime numbers, and even patterns that self-replicate. In fact, this humble game with two simple rules is capable of carrying out *any* computer algorithm, which in jargon means it is "Turing complete," which is utterly mind-blowing.

Now that we have the basics, we can get to the analogy that might make reading about grid cells worth it. Let's pretend for a moment that the Game of Life grid represents an actual universe, albeit one with two dimensions and on a grid. In this universe there are the two simple laws of nature that govern everything that can happen. We have the benefit of being external omniscient scientists who can see this universe evolve from the third dimension, and we are trying to determine the laws of nature in this grid universe.

If you didn't know the rules of the game, but you could play around with it, I am reasonably confident that you could eventually determine the rules through trial and error. You would perhaps even gain enough confidence to call them "laws" after you had done due diligence testing everything you could test. However, any beings living in this two-dimensional universe would probably be pretty confused about what was going on, with their friends and neighbors just popping in and out of existence capriciously, so what if we just stick with observing?

If you can't manipulate the grid universe to test your hypotheses, the rules you might be able to infer would be highly dependent on how the grid was populated to begin with. For example, take one of the three simple patterns we started with; if this grid only started with single isolated pixels being populated, you might infer a rule for the universe along the lines of "Everything dies after one time step and nothing else happens," and that is just a really sucky universe. With the blinker, you could infer that only shapes that are three pixels long exist and they alternate in direction; so at least something is happening in the universe, but it is a far cry from being rich with complexity. Finally, the blocks just get us nowhere—in this case, "Blocks exist" is about the best rule you could come up with. On the other hand, if this grid started with complex patterns and you could watch them unfold and keep track of what happened (i.e., collect data), you might be able to come up with more sophisticated hypotheses, and perhaps eventually even call them "theories" or "laws."

At this point, as a two-dimensional-grid-universe external observer, you still have the leisure of watching individual time steps unfold and the luxury of identifying individual pixels as they turn on and off. What if, just to make this more realistic, we sped up the frame rate and zoomed way out so that both individual time steps and pixels were blurred—in fact, as an observer you don't even *know* there are individual pixels at the bottom of this behavior (though you might hypothesize as much). Could you determine the rules? Perhaps, but not easily, and probably not with a high degree of confidence. The situation may be starting to seem quite futile. But it gets worse.

Not only are we mostly relegated to passively observing this universe, and the time and size scales are blurred to obscurity, to bring this analogy onto the actual universe we live in, we have to be *a part* of it. So now imagine you are a scientist *inside* this two-dimensional universe trying to understand how it works, which means you don't have the ability to see the grid from above. For things that are extraordinarily nearby you can perhaps walk (or more likely slide) around them to get a better sense of their behavior. But for things outside of your provincial local environment, you can only see things in projection, and you only see cells within your horizon. Blinkers would appear to alternate between one cell and three cells. Blocks would look like lines of two cells. Complex patterns unfolding might appear totally random. Could you determine the laws of nature in this simple universe? Possibly. But it would take extraordinary patience and abstract thinking.

The point of this thought experiment is that this fictitious state—trying to understand even the two simple laws of the grid universe—is basically our situation in the universe in which we actually live. We have (relatively) terrible spatial and temporal resolution. There isn't much we can manipulate, so we are stuck passively observing whatever the universe makes available to us. *And* we are part of this universe.

To be sure, we have laboratory experiments that attempt to manipulate whatever things we can manipulate to test our hypotheses, but these attempts are extremely limited in the physical conditions they can probe; for example, we are many orders of magnitude away from investigating energy levels at which the four forces may be unified. As a result, we are mostly restricted to simply observing whatever we can to look for patterns in behavior (and this is nearly 100 percent true for astronomers) to sniff out clues. Blurring out the individual time steps and pixels starts to recreate our attempts to understand what is happening in the quantum realm; the reality on quantum scales of space and time is entirely blurred to our perception and experiments. Perhaps the Planck length is the actual size of "pixels" in our universe, and the time steps between iterations happen on the Planck time. Yet we persist.

Back in our actual universe, we have a challenge before us to try to determine the underlying laws of nature. Let's start by trying to map out the part of the landscape that we can and distill what we think we know. We need to sort through some semantics here, which I know makes for riveting reading. But if we don't agree on what we mean by the word "law," this chapter is just not going to go well.

There are several ways one might use the word "law" in the context of physics (we won't even touch the legal system). For our purposes here, it is important to clearly distinguish between a "scientific law" and a "law of nature." A scientific law refers to the use of the word "law" in the scientific method, as you read in the chapter on epistemology. In this case, we mean a prescription that has been well tested but could potentially be falsified. As a reminder, even within this context, science does not use the word "law" self-consistently: there are things that have "law" in their name that are not actually laws (e.g., Newton's law of gravity), and there are things that are not called laws that probably should be (e.g., the theory of general relativity). We generally have high confidence in this proximate type of "law," but we also hold our positions to be defeasible—if there were sufficient evidence to the contrary, we can and should change our minds (which is, in fact, what caused Newton's "law" of gravity to be superseded).

Ultimately, the laws we infer could just be a projection or manifestation of a deeper law. In this scenario, a positivist might argue that, if a proposed law perfectly describes the observations, the model is interchangeable with reality. I would argue that the positivist position has some practical value, but we risk misleading ourselves and missing out on deeper truths. This deeper truth is what I am calling a true law of nature.

The true underlying laws of nature are what we really want to know, but they may be beyond our ability to grasp or test. This version of a "law" has a bit of a Platonic flavor—it is the sort of ideal law that might reside in Plato's realm of forms, in which the very essence of a thing is said to exist. Sometimes when I am teaching this subject material, I will have my students read Plato's *Allegory of the Cave* (which you are, of course, welcome to do if you want extra credit). The laws we infer could be only a shadow of the true underlying form of the law.

The point of dragging you through these two types of laws is to appreciate the difference between what we can do and understand as mere mortals and what may be a more fundamental reality we are trying to uncover. The laws of the second category, the true underlying fundamental laws of nature, are what we are after in this chapter.

These fundamental laws necessarily have a number of character-istics: they are omnipotent, meaning everything in the universe (or multiverse) is subject to them; they are universal (or multiversal), meaning they apply everywhere in all dimensionality that exists; they are atemporal, meaning they are true at all times (if they were not true at all times, that would suggest the existence of a more funda-mental law that dictates how they can change); and they are absolute, meaning they are not contingent on anything else (if they depended on something else, that would suggest a more fundamental underly-ing law).

The situation gets metaphysically muddy at this point. In fact, there is a philosophical debate as to whether the laws of nature are "necessary," which in the philosophical usage of the word means "the laws could not have been otherwise." If these laws were *con-tingent* on some other factor, we are left wondering what that other factor is, which must be a still deeper underlying requirement. To be clear about my own bias, I mostly fall into the metaphysical philos-ophy camp[3] of "necessitarians," which means I lean toward a belief that there are "necessary" laws (or perhaps even a single law) that govern all of reality. This is a core axiom I take as given, at least in part because I simply can't fathom a logical scheme in which there are not fundamental laws from which behaviors in the cosmos arise. That being said, just because something doesn't make logical sense to me, doesn't mean it might not be true. In fact, in debating whether the laws of nature are "necessary," we run smack into both issues of infinite regression and causality yet again. I'm sure you're shocked.

In the case of laws, we could imagine that there are some set of necessary laws from which the rest of the universe (or multiverse) takes its instructions. If these laws are philosophically necessary, they *could not have been* and *could not be otherwise.* The first ques-tion that might surface in your mind might be along the lines of

"Well, where did those laws come from, and why those particular laws and not some other laws?" Good questions. That is the problem with being "necessary"—it is equivalent to saying, "That is just the way things are," which is deeply unsatisfying. A second issue with there being necessary laws—that *could not have been otherwise*—is the challenge in figuring out how anything contingent could ever happen without being necessary itself: if the laws are necessary and could not be otherwise, then what arises from those laws could also not be otherwise. In the end, we end up with a universe (or multiverse) that appears to be entirely "necessary," which—for example—leaves no room for beloved notions like free will. On the other hand, if the laws are contingent, what are they contingent on? And around and around we go. There is clearly more to the story.

It is not hard to understand why countless theologians and philosophers have turned toward these ontological arguments as proof of God, including a formal logical argument by famed logician and mathematician Kurt Gödel (whom we will see again later in this chapter).[4] My own thinking on this is that our apparent inability to untangle the philosophical knot of whether the laws of nature are necessary *does not prove* the existence of a higher power or intelligence—to my mind, this is just a "God of the gaps" fallacy (see Chapter 2) wearing a high-end cloak of philosophy. However, I think this philosophical conundrum *does prove* there is an aspect of the cosmos (which may or may not include a higher power) that defies current human logic and comprehension. This should not really be a surprise if we have an ounce of humility about our own intellectual ability—just because we might be the most intelligent species on this planet does not imply we represent the pinnacle of intelligence, nor that we are able to understand everything about reality, try as we might.

This is one of those times when the most productive path forward might be to acknowledge that there is a mystery lurking about and start by taking stock of what we think we know about how the laws of nature are manifest in our universe and see if it gives us any insight.

The Four Known Forces

We tend to think of the laws of nature and the forces that act in the universe as synonymous, but this need not be the case. Rather, the forces influence how things behave through interactions. In turn, the characteristics of these interactions are manifestations of the laws of nature made visible to us through the properties of the universe we happen to live in. This may sound like a picky semantics point, but distinguishing between the observable forces and the underlying laws is important when we consider why we have the forces that we do (and whether they could have been otherwise).

That being said, the known forces in the universe are the only handle we currently have on grasping any underlying laws, so trying to understand these forces and their properties is essential in getting to the bottom of the mystery at hand. We currently have four known forces that operate in our universe (if you want to sound more physics-y, you can refer to these as "fundamental interactions"): the gravitational force, the electromagnetic force, the weak force, and the strong force. You might have noted the word "currently" in the previous sentence—we have only known about the weak and strong interactions for less than a century, so I think we would be remiss to exclude the possibility that we might discover something new.[5] These four forces appear to be radically different, which doesn't sit well in modern physics because physicists really like symmetry. Many of us believe, with almost religious fervor, that elegant symmetries must underlie the architecture of the universe.

There are many reasons for this conviction, some of which lean into aesthetics—nonsymmetries are messy, and messy requires more ad hoc explanations, and ad hoc explanations feel philosophically uncomfortable. Beyond the aesthetics of symmetry, history has also been a guide; in building our understanding of the universe, we have been shown over and over again that, all things being equal, the default of nature is symmetry. Generally, physicists don't need an explanation when things are symmetric, but when things are *not* symmetric there darn well better be a reason.

To illustrate one aspect of the apparent asymmetry, I want you to imagine a little demonstration, which you can easily carry out on your own (if you can't, that's OK—I am confident that you will be able to follow along). When I do this demonstration in class, I tell my students this is the most mind-blowing demonstration I can do the entire semester. Here is what you need to do: find any old magnet—this could just be a little refrigerator magnet, or something like the miniature magnetic calendars I get from my insurance salesman every year (does anyone use these as calendars?). In fact, the smaller the magnet, the better, because that will make this demonstration all the more impressive. Next, find something small and magnetic, like a paper clip. Now carefully hold the magnet and attach the paper clip to dangle from underneath it. Ta-da!

Do you see how extraordinary this is? If you are like my students, at this point there is usually some polite or nervous laughter and odd looks suggesting they are questioning their curricular choices. The *entire mass of the Earth* is pulling down on that paper clip (or whatever little object you used), but this tiny little throwaway magnet has enough force to counteract *the entire planet*. If you have gone your whole life without appreciating how ridiculous this imbalance is, I promise you are not alone. Yet, this simple do-it-yourself experiment reveals one of the great mysteries of modern physics—the fact that gravity is unfathomably weak compared to the other forces. Even the weak force (which comes by its name honestly) is roughly 10^{33} times stronger than gravity.

I hope you are curious to know a little more about the forces now. The four forces each have their own personality (envision an online quiz entitled "Which of the Four Forces Are You?"). The key characteristics of the forces that distinguish them from one another are their range (i.e., how far away they can affect things); their strength (in basic physics we parametrize these strengths as things called "coupling constants"); the characteristics the forces couple to (I think of this as each force having its own kind of Velcro, which I realize makes no sense at all, but nevertheless that is the analogy I've had in my mind since I was in college); and the type of particle the force uses to interact (these are "virtual" particles, also known as "exchange" particles). I am sure that a clever person can map those onto human characteristics, and possibly even put them into a dating profile.

Gravity has an outsized role in our lives, at least in part because we are aware of our continuous interactions with it. I will admit that I go through my day and mostly don't even think about gravity; it is just there, completely in the background, which just doesn't do justice to how weird it is. Gravity comes from mass, but have you ever thought about what mass even is? This is one of those multitude of things I think most of us go through life taking for granted because it is so utterly familiar. Many of my students understandably have trouble getting their minds around how a particle—or any *thing*—could not have mass. In our normal lives we are duped into thinking that "thingness" goes hand in hand with having mass. But in the world of particle physics, this is not so. Just like a particle may or may not have charge, or a pizza may or may not have pepperoni, a particle may or may not have mass.

In fact, in our current understanding of mass, particles that interact with gravity don't inherently have mass at all—rather mass is only acquired by certain types of particles when they interact with a field called the "Higgs field" (the discovery of which won the Nobel Prize in 2013). If we really want to stretch the pizza analogy where it was never intended to go, pizzas do not inherently have pepperoni, but some pizzas acquire pepperoni in the assembly line. Vegetarian pizzas, as a rule, do not. So, in this pizza-particle world, particles like photons would be pepperoni-free, but they are still pizzas.

If I anthropomorphize the forces, I think of gravity as the little force with a chip on its shoulder and something to prove. It probably drives a pickup around town with more horsepower than it will ever need. If you list the forces in order of strength, gravity is at the bottom—as we've already illustrated with a handy magnet and a paper clip, gravity is mind-blowingly weak.

The two reasons gravity takes on this outsized role in our lives are (1) that it has an infinite range, and (2) that it doesn't get canceled out by a negative gravity (at least not apparently in our normal daily lives, but remember dark energy). The reason gravity has an infinite range is that the hypothetical exchange particle that carries it, the graviton, doesn't have any mass, so it gets to travel off into the universe as far as it wants to. The more massive the carrier particle, the shorter the range. There is this tiny issue that we haven't detected a graviton yet, but let's just sweep that under the rug for now.

In contrast to gravity (which is easy to take for granted), when I play with magnets it seems like sorcery, with forces that can't be seen or touched. But at least we can interact with electromagnetism in our daily lives, which gives us some sense of its reality and behavior. As we saw with the paper clip and magnet above, electromagnetism is *way* stronger than gravity, by a factor of around 10^{33}. True to its name, this force comes about from interactions between electromagnetic fields and particles with charge. Electricity and magnetism were originally thought to be two separate forces but were merged into a single force in 1873 by James Clerk Maxwell (after whom Maxwell's equations are appropriately named).[6] If you are wondering what electrically charged particles and magnetism have in common, when electric charges move, they generate a magnetic field, so the two concepts are tightly entangled.

You may be wondering how the magnet you used (or envisioned) above has a magnetic field if there are no charges moving around, which is a great question; the answer comes down to the minuscule little electrons that are in the atoms that make up the magnet—these little fellows are spinning, and because they spin, they have something we call a magnetic dipole, which is where the magnetic field comes from (once again, I want to pause and thank whoever called it a "magnetic dipole" for giving it a name that actually makes sense). The reason a magnet is magnetized (as opposed to, say, an apple) is that in a permanent magnet like my miniature insurance calendar, the magnetic dipoles more or less line up and face the same direction. The electrons in an apple have magnetic dipoles as well, but they are oriented every which way and mostly just cancel each other out. Most everyday objects have a bit of a net magnetic field—a disturbing example of which can be seen in a physics experiment that levitated a frog using magnetic fields[7] (which you can easily find by a quick search of "levitating frog"). I don't think anyone has tried this with humans yet, and that is almost certainly a good thing.

Like the gravitational force, the electromagnetic force has an infinite range, in this case due to its carrier particle, the photon, not having any mass. In contrast to the graviton, we are pretty sure that photons exist. Given that both gravity and electromagnetism have infinite ranges, electromagnetism would overwhelmingly dominate our lives if not for the simple fact that positive and negative charges can cancel each other out. The fact that there are both positive and negative charges and that they usually almost completely balance each other is an astounding fact hiding in plain sight, especially in contrast with gravity (which always seems to be the odd duck among the forces).

The strong and weak forces are another matter entirely.[8] If you routinely perceive interactions with the weak or strong force as part of your normal life, probably you should seek professional help. The exact behavior of the strong force is important when we talk about fine-tuning in the next chapter, so I want you to make a mental note of all the different knobs and dials that could be tweaked if you were an omniscient being designing a force. If you were trying to write a novel with quarks and gluons as characters, it would be very hard to keep track of who was doing what, and your editor would probably tell you to ease off on the number of characters. For better or worse, the universe did not have an editor with this requirement, which makes for a very thick plot. If your eyes glaze over the next couple paragraphs, that is OK, but know that you can come back to them later if you want to. The point is to illustrate how many different parameters are at play.

The strong force is more complicated than the gravitational and electromagnetic forces, and also a fair bit less intuitive, so I'm going to skip over a lot of the physics for the sake of not completely changing the title of this book to *Introduction to Nuclear Physics*. The strong interaction works to keep subatomic particles called "quarks" close together (usually in pairs or triplets), using particles that carry the force called "gluons." This is what keeps nuclei of atoms together; without the strong force, the mutual electromagnetic repulsion of protons would keep them apart and we wouldn't have any atoms other than hydrogen.

Both gluons and quarks have their own characteristics, which include a "color"; you can kind of think of "color" in this context as being like an electric charge in the sense that it is a property of the sub-atomic particle and determines who it is allowed to be friends with in the particle-physics novel. But you should not think of them as actually having a color, although this whimsical notion can help to visualize them and keep them straight. When you mix three "primary"-colored quarks together, the resulting particle becomes "colorless" (i.e., it has a net color of 0), which means it is allowed to exist by itself, like an adult child who has their own bank account, insurance, and place to live. In case that isn't complicated enough, in addition to the color charges, there are six separate types of quarks, which are each said to have a "flavor" (please grant physicists some poetic license here for their use of the word "flavor," and maybe also don't let them season your food): up, down, charm, strange, top, and bottom.

Like the electromagnetic force, the strong interaction can be either positive or negative, but a key difference from electromagnetism is that the sign of this force doesn't depend on the particles themselves, but rather it depends on the *distance* between them—at small distances the strong force becomes extremely repulsive (which is good, otherwise nuclei of atoms would collapse); this repulsion is ultimately responsible for the size of nuclei. I think of the strong force as being a bit like a squishy rubber ball; the more you stretch it, the more it pulls in (attraction), and the more you squeeze it, the more it pushes back (repulsion); there is a "natural" length scale that it prefers to be. You might wonder why that particular length scale is preferred (so do I), but it is fair to say we don't yet know.

At the end of the day, you can thank the strong force for the fact that we have atoms at all, without which your daily life would be radically different, and you probably would not be thinking about physics (or anything else).

Finally, we get to the weak force, which I've saved until last for two reasons: first, because I think it is the least intuitive, and second, because I think it is the most fascinating and may hold the keys to some of the mysteries in the universe we are wrestling with. The weak force is aptly named, given that it is several orders of magnitude weaker than the strong or electromagnetic force (yet let us recall that gravity is 10^{33} times weaker still). Because the carrier particles for the weak force (known as the "W+, W−, and Z bosons," which are names perhaps only a parent could love) are extremely massive—even more massive than a proton or neutron—the weak force has a very short range, which is one of the reasons why it is so elusive.

To understand the weak force, it might be helpful to note that what we perceive as forces are—at their core—an exchange of virtual particles (hence the preferred use of the word "interaction" instead of "force"—see note 8 for this chapter). This virtual particle exchange doesn't always need to just "push" or "pull"; it can have other effects, too. By analogy, think about holding hands with someone; you could pull them closer or keep them at arm's length (i.e., pull or push), but you could also warm their hand or make them feel happy or self-conscious.

The weak interaction does all of this. In particular, one of the things the weak interaction can do is make the other particles feel self-conscious enough that they *change their identity*, which in turn can lead to radioactive decay. For example, sitting in the nucleus of an atom is a proton, which is made of three quarks, which each have their own "flavor" (two "up" quarks and one "down" quark). If the proton is able to capture an electron, the weak force can cause one of the quarks to change their flavor from "up" to "down," which turns the proton into a neutron, which causes the particle to decay.[9] We should all be extremely grateful for this odd characteristic of the weak force, because this decay of a proton into a neutron is what enables nuclear fusion in the center of stars. In fact, given that we define stars as objects that have nuclear fusion in their cores, we wouldn't have stars *at all* if not for the weak force.

The fact that the weak force can cause subatomic particles to change their identity is fascinating enough, but the weak force is a decisively complex character with more to offer our plot. Do you remember when we talked about the arrow of time and the asymmetry between matter and antimatter? Both of these have potential ties to charge-parity (CP) symmetry breaking, and the weak force is the *only* thing in physics known to violate CP symmetry. The upshot of all of this is that this bizarre and unassuming weak interaction is extraordinarily important in the universe as we know it. Moreover, our very existence is dependent on there being asymmetries in the universe—like CP violation and the asymmetry between matter and antimatter—yet we generally believe that symmetry is the natural state. Something has to give.

Symmetry and Unification?

I will confess that I am enchanted by symmetries, and I am prone to see them everywhere in the natural world, which can be really annoying to my companions when out for a hike. I even occasionally teach a course entitled "Beauty and Math in the Cosmos," in which the symmetries of the universe have a whole unit devoted to them.[10] When I encounter broken symmetries, I am equally curious—because these broken symmetries signal that *something* happened to cause the symmetries to break. Take the human body for example, which typically has a very high degree of reflection symmetry about the vertical axis. Yet, the heart is normally extended toward the left side,[11] and most people have a dominant hand. Even faces are not perfectly symmetric. In each of these cases, asking "why" leads to *something* that caused an asymmetry. In the case of the human heart, the asymmetry seems to be due to the left ventricle needing to be larger, which then leads to the question of why the left ventricle is larger. Or with handedness, the preference appears to be linked to how the brain organizes functions, which then leads to the question of why that organization has an asymmetry.

Yet, if the universe were entirely symmetric, nothing of note would exist—the only way to exist in perfect symmetry (reflection, translation, rotation, etc.) is to have a perfectly uniform or empty existence. So, the asymmetries in our universe are profoundly important. The fact that our universe appears to be in an intermediate state between having fundamental symmetries *and* asymmetries is both profoundly important and intellectually perplexing.

Given the apparent disposition of the universe, physicists spend a lot of time thinking about why we have symmetries, and perhaps even more time thinking about why symmetries don't always hold. There is a broad inclination to conjecture that the four forces (as they are manifest today) are not actually different forces, but rather perhaps aspects of a single force that—for some reason—had its symmetry broken. What does that even mean? I realize that we are taking concepts that are already abstract, then shaking them up, spinning them around, and looking at them while standing on our heads. So, here is an analogy: imagine a shape like the following:

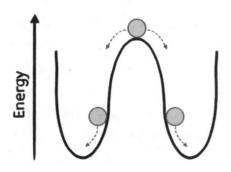

In this diagram, a ball initially sits at the top of the curve in the middle. When the ball is in this initial position, the entire system is symmetric. But if something happens—perhaps a brief gust of wind—the ball will roll to one side or the other, resulting in the symmetry being broken. As in this diagram, symmetry breaking in physics generally entails going from a higher energy symmetric state to a lower energy asymmetric state. This is more or less what we mean when we refer to a broken symmetry, and also why we expect more symmetries to emerge at higher energy levels.

The *cause* of the ball toppling over is a different matter, given that gusts of wind aren't typically at play in the fundamental physics world. Often this symmetry breaking is referred to as "spontaneous," which is kind of code for "We don't really have a reason, but we know that it happens." Insofar as we have a reason, it often invokes quantum mechanics, which can (and does) cause things to wiggle around and tunnel between different states of being.

When it comes to the four known forces, the idea is that these forces are really just the lower energy asymmetries of what would appear as a single force at a higher energy. In other words, each of these forces is akin to that ball in the above diagram falling to the left or the right. But if we could dial up the energy level, these forces would become increasingly unified into a single symmetric state with the ball perched precariously in the middle.

To be sure, physicists aren't just being capricious in the expectation of symmetry with these four fundamental forces, rather, there is solid precedent. First, we realized that electricity, magnetism, and light were all aspects of the same force, thanks to Maxwell way back in 1873 (it's not too late to get your tattoo). Then almost a century later, we were able to unify electromagnetism and the weak interaction into the electroweak interaction (which won the Nobel Prize in 1979).[12] Since the unification of the weak and electromagnetic forces, there has been a broad expectation that the strong interaction would join the party at even higher energies—by which I mean energies around 10^{16} GeV, which is roughly 1,000 times higher than we can currently reach with the Large Hadron Collider. New colliders on the horizon will "only" reach energies roughly on order of magnitude higher than the Large Hadron Collider. The good news is that I may not need to revise this section of the book anytime soon. This hypothetical unification of the electroweak and strong interactions is referred to as the "grand unified theory," which tells me that the folks who came up with this name did not care about winsome acronyms (aka GUT).

In the meantime, our best theory to understand the zoo of particles and their interactions is the "standard model" (kudos for creativity on the name), which does a decent job of explaining three of the four forces. Except gravity. To give you a visual sense of the level of "simplicity" in the standard model, it takes a full page in the book to write it out. Not to pass aesthetic judgment, but to my eye this equation does not look particularly elegant and symmetric. In contrast to Maxwell's equations, I don't know a single physicist who has this tattooed on themselves.

A notable feature of the unification scheme we have in mind is that the time at which we think the universe reached GUT energies corresponds to the time during which we think cosmic inflation happened. This correspondence smacks of more than a random coincidence. In fact, the working hypothesis is that when the strong and electroweak interactions were rendered apart, the fracturing that broke the symmetry injected an unfathomable amount of energy into the very fabric of space-time, and this influx of energy is a leading contender for being at the root of cosmic inflation.

For better or worse, I default to visualizing this fracturing of forces as similar to (albeit a dash more powerful) a spontaneous

$$\mathcal{L}_{SM} = -\tfrac{1}{2}\partial_\nu g^a_\mu \partial_\nu g^a_\mu - g_s f^{abc}\partial_\mu g^a_\nu g^b_\mu g^c_\nu - \tfrac{1}{4}g_s^2 f^{abc}f^{ade}g^b_\mu g^c_\nu g^d_\mu g^e_\nu - \partial_\nu W^+_\mu \partial_\nu W^-_\mu -$$
$$M^2 W^+_\mu W^-_\mu - \tfrac{1}{2}\partial_\nu Z^0_\mu \partial_\nu Z^0_\mu - \tfrac{1}{2c_w^2}M^2 Z^0_\mu Z^0_\mu - \tfrac{1}{2}\partial_\mu A_\nu \partial_\mu A_\nu - igc_w(\partial_\nu Z^0_\mu(W^+_\mu W^-_\nu -$$
$$W^+_\nu W^-_\mu) - Z^0_\nu(W^+_\mu \partial_\nu W^-_\mu - W^-_\mu \partial_\nu W^+_\mu) + Z^0_\mu(W^+_\nu \partial_\nu W^-_\mu - W^-_\nu \partial_\nu W^+_\mu)) -$$
$$igs_w(\partial_\nu A_\mu(W^+_\mu W^-_\nu - W^+_\nu W^-_\mu) - A_\nu(W^+_\mu \partial_\nu W^-_\mu - W^-_\mu \partial_\nu W^+_\mu) + A_\mu(W^+_\nu \partial_\nu W^-_\mu -$$
$$W^-_\nu \partial_\nu W^+_\mu)) - \tfrac{1}{2}g^2 W^+_\mu W^-_\mu W^+_\nu W^-_\nu + \tfrac{1}{2}g^2 W^+_\mu W^-_\nu W^+_\mu W^-_\nu + g^2 c_w^2(Z^0_\mu W^+_\mu Z^0_\nu W^-_\nu -$$
$$Z^0_\mu Z^0_\mu W^+_\nu W^-_\nu) + g^2 s_w^2(A_\mu W^+_\mu A_\nu W^-_\nu - A_\mu A_\mu W^+_\nu W^-_\nu) + g^2 s_w c_w(A_\mu Z^0_\nu(W^+_\mu W^-_\nu -$$
$$W^+_\nu W^-_\mu) - 2A_\mu Z^0_\mu W^+_\nu W^-_\nu) - \tfrac{1}{2}\partial_\mu H\partial_\mu H - 2M^2\alpha_h H^2 - \partial_\mu \phi^+\partial_\mu \phi^- - \tfrac{1}{2}\partial_\mu \phi^0\partial_\mu \phi^0 -$$
$$\beta_h\left(\tfrac{2M^2}{g^2} + \tfrac{2M}{g}H + \tfrac{1}{2}(H^2 + \phi^0\phi^0 + 2\phi^+\phi^-)\right) + \tfrac{2M^4}{g^2}\alpha_h -$$
$$g\alpha_h M\left(H^3 + H\phi^0\phi^0 + 2H\phi^+\phi^-\right) -$$
$$\tfrac{1}{8}g^2\alpha_h\left(H^4 + (\phi^0)^4 + 4(\phi^+\phi^-)^2 + 4(\phi^0)^2\phi^+\phi^- + 4H^2\phi^+\phi^- + 2(\phi^0)^2 H^2\right) -$$
$$gMW^+_\mu W^-_\mu H - \tfrac{1}{2}g\tfrac{M}{c_w^2}Z^0_\mu Z^0_\mu H -$$
$$\tfrac{1}{2}ig\left(W^+_\mu(\phi^0\partial_\mu \phi^- - \phi^-\partial_\mu \phi^0) - W^-_\mu(\phi^0\partial_\mu \phi^+ - \phi^+\partial_\mu \phi^0)\right) +$$
$$\tfrac{1}{2}g\left(W^+_\mu(H\partial_\mu \phi^- - \phi^-\partial_\mu H) + W^-_\mu(H\partial_\mu \phi^+ - \phi^+\partial_\mu H)\right) + \tfrac{1}{2}g\tfrac{1}{c_w}(Z^0_\mu(H\partial_\mu \phi^0 - \phi^0\partial_\mu H) +$$
$$M(\tfrac{1}{c_w}Z^0_\mu \partial_\mu \phi^0 + W^+_\mu \partial_\mu \phi^- + W^-_\mu \partial_\mu \phi^+) - ig\tfrac{s_w^2}{c_w}MZ^0_\mu(W^+_\mu \phi^- - W^-_\mu \phi^+) + igs_w MA_\mu(W^+_\mu \phi^- -$$
$$W^-_\mu \phi^+) - ig\tfrac{1-2c_w^2}{2c_w}Z^0_\mu(\phi^+\partial_\mu \phi^- - \phi^-\partial_\mu \phi^+) + igs_w A_\mu(\phi^+\partial_\mu \phi^- - \phi^-\partial_\mu \phi^+) -$$
$$\tfrac{1}{4}g^2 W^+_\mu W^-_\mu(H^2 + (\phi^0)^2 + 2\phi^+\phi^-) - \tfrac{1}{8}g^2\tfrac{1}{c_w^2}Z^0_\mu Z^0_\mu(H^2 + (\phi^0)^2 + 2(2s_w^2 - 1)^2\phi^+\phi^-) -$$
$$\tfrac{1}{2}g^2\tfrac{s_w^2}{c_w}Z^0_\mu \phi^0(W^+_\mu \phi^- + W^-_\mu \phi^+) - \tfrac{1}{2}ig^2\tfrac{s_w^2}{c_w}Z^0_\mu H(W^+_\mu \phi^- - W^-_\mu \phi^+) + \tfrac{1}{2}g^2 s_w A_\mu \phi^0(W^+_\mu \phi^- +$$
$$W^-_\mu \phi^+) + \tfrac{1}{2}ig^2 s_w A_\mu H(W^+_\mu \phi^- - W^-_\mu \phi^+) - g^2\tfrac{s_w}{c_w}(2c_w^2 - 1)Z^0_\mu A_\mu \phi^+\phi^- -$$
$$g^2 s_w^2 A_\mu A_\mu \phi^+\phi^- + \tfrac{1}{2}ig_s \lambda^a_{ij}(\bar{q}^\sigma_i \gamma^\mu q^\sigma_j)g^a_\mu - \bar{e}^\lambda(\gamma\partial + m^\lambda_e)e^\lambda - \bar{\nu}^\lambda(\gamma\partial + m^\lambda_\nu)\nu^\lambda - \bar{u}^\lambda_j(\gamma\partial +$$
$$m^\lambda_u)u^\lambda_j - \bar{d}^\lambda_j(\gamma\partial + m^\lambda_d)d^\lambda_j + igs_w A_\mu\left(-(\bar{e}^\lambda\gamma^\mu e^\lambda) + \tfrac{2}{3}(\bar{u}^\lambda_j\gamma^\mu u^\lambda_j) - \tfrac{1}{3}(\bar{d}^\lambda_j\gamma^\mu d^\lambda_j)\right) +$$
$$\tfrac{ig}{4c_w}Z^0_\mu\{(\bar{\nu}^\lambda\gamma^\mu(1 + \gamma^5)\nu^\lambda) + (\bar{e}^\lambda\gamma^\mu(4s_w^2 - 1 - \gamma^5)e^\lambda) + (\bar{d}^\lambda_j\gamma^\mu(\tfrac{4}{3}s_w^2 - 1 - \gamma^5)d^\lambda_j) +$$
$$(\bar{u}^\lambda_j\gamma^\mu(1 - \tfrac{8}{3}s_w^2 + \gamma^5)u^\lambda_j)\} + \tfrac{ig}{2\sqrt{2}}W^+_\mu\left((\bar{\nu}^\lambda\gamma^\mu(1 + \gamma^5)U^{lep}_{\lambda\kappa}e^\kappa) + (\bar{u}^\lambda_j\gamma^\mu(1 + \gamma^5)C_{\lambda\kappa}d^\kappa_j)\right) +$$
$$\tfrac{ig}{2\sqrt{2}}W^-_\mu\left((\bar{e}^\kappa U^{lep\dagger}_{\kappa\lambda}\gamma^\mu(1 + \gamma^5)\nu^\lambda) + (\bar{d}^\kappa_j C^\dagger_{\kappa\lambda}\gamma^\mu(1 + \gamma^5)u^\lambda_j)\right) +$$
$$\tfrac{ig}{2M\sqrt{2}}\phi^+\left(-m^\kappa_e(\bar{\nu}^\lambda U^{lep}_{\lambda\kappa}(1 - \gamma^5)e^\kappa) + m^\lambda_\nu(\bar{\nu}^\lambda U^{lep}_{\lambda\kappa}(1 + \gamma^5)e^\kappa) +$$
$$\tfrac{ig}{2M\sqrt{2}}\phi^-\left(m^\lambda_e(\bar{e}^\lambda U^{lep\dagger}_{\lambda\kappa}(1 + \gamma^5)\nu^\kappa) - m^\kappa_\nu(\bar{e}^\lambda U^{lep\dagger}_{\lambda\kappa}(1 - \gamma^5)\nu^\kappa) - \tfrac{g}{2}\tfrac{m^\lambda_\nu}{M}H(\bar{\nu}^\lambda\nu^\lambda) -$$
$$\tfrac{g}{2}\tfrac{m^\lambda_e}{M}H(\bar{e}^\lambda e^\lambda) + \tfrac{ig}{2}\tfrac{m^\lambda_\nu}{M}\phi^0(\bar{\nu}^\lambda\gamma^5\nu^\lambda) - \tfrac{ig}{2}\tfrac{m^\lambda_e}{M}\phi^0(\bar{e}^\lambda\gamma^5 e^\lambda) - \tfrac{1}{4}\bar{\nu}_\lambda M^R_{\lambda\kappa}(1 - \gamma_5)\hat{\nu}_\kappa -$$
$$\tfrac{1}{4}\bar{\nu}_\lambda M^R_{\lambda\kappa}(1 - \gamma_5)\hat{\nu}_\kappa + \tfrac{ig}{2M\sqrt{2}}\phi^+\left(-m^\kappa_d(\bar{u}^\lambda_j C_{\lambda\kappa}(1 - \gamma^5)d^\kappa_j) + m^\lambda_u(\bar{u}^\lambda_j C_{\lambda\kappa}(1 + \gamma^5)d^\kappa_j) +$$
$$\tfrac{ig}{2M\sqrt{2}}\phi^-\left(m^\lambda_d(\bar{d}^\lambda_j C^\dagger_{\lambda\kappa}(1 + \gamma^5)u^\kappa_j) - m^\kappa_u(\bar{d}^\lambda_j C^\dagger_{\lambda\kappa}(1 - \gamma^5)u^\kappa_j) - \tfrac{g}{2}\tfrac{m^\lambda_u}{M}H(\bar{u}^\lambda_j u^\lambda_j) -$$
$$\tfrac{g}{2}\tfrac{m^\lambda_d}{M}H(\bar{d}^\lambda_j d^\lambda_j) + \tfrac{ig}{2}\tfrac{m^\lambda_u}{M}\phi^0(\bar{u}^\lambda_j\gamma^5 u^\lambda_j) - \tfrac{ig}{2}\tfrac{m^\lambda_d}{M}\phi^0(\bar{d}^\lambda_j\gamma^5 d^\lambda_j) + \bar{G}^a\partial^2 G^a + g_s f^{abc}\partial_\mu\bar{G}^a G^b g^c_\mu +$$
$$\bar{X}^+(\partial^2 - M^2)X^+ + \bar{X}^-(\partial^2 - M^2)X^- + \bar{X}^0(\partial^2 - \tfrac{M^2}{c_w^2})X^0 + \bar{Y}\partial^2 Y + igc_w W^+_\mu(\partial_\mu\bar{X}^0 X^- -$$
$$\partial_\mu\bar{X}^+ X^0) + igs_w W^+_\mu(\partial_\mu\bar{Y}X^- - \partial_\mu\bar{X}^+ Y) + igc_w W^-_\mu(\partial_\mu\bar{X}^- X^0 -$$
$$\partial_\mu\bar{X}^0 X^+) + igs_w W^-_\mu(\partial_\mu\bar{X}^- Y - \partial_\mu\bar{Y}X^+) + igc_w Z^0_\mu(\partial_\mu\bar{X}^+ X^+ -$$
$$\partial_\mu\bar{X}^- X^-) + igs_w A_\mu(\partial_\mu\bar{X}^+ X^+ -$$
$$\partial_\mu\bar{X}^- X^-) - \tfrac{1}{2}gM\left(\bar{X}^+ X^+ H + \bar{X}^- X^- H + \tfrac{1}{c_w^2}\bar{X}^0 X^0 H\right) + \tfrac{1-2c_w^2}{2c_w}igM\left(\bar{X}^+ X^0\phi^+ - \bar{X}^- X^0\phi^-\right) -$$
$$\tfrac{1}{2c_w}igM\left(\bar{X}^0 X^-\phi^+ - \bar{X}^0 X^+\phi^-\right) + igMs_w\left(\bar{X}^0 X^-\phi^+ - \bar{X}^0 X^+\phi^-\right) +$$
$$\tfrac{1}{2}igM\left(\bar{X}^+ X^+\phi^0 - \bar{X}^- X^-\phi^0\right).$$

This is what the standard model of particle physics looks like at the moment. Even the uninitiated can see that this equation (or Lagrangian in physics terms) does not seem particularly elegant or symmetric. Credit: Matilde Marcolli, who deserves an award for having the patience to carefully type this in.

fracturing of a thick ice sheet covering a frozen lake. This particular visualization is likely a product of growing up in Minnesota and being intimately familiar with frozen lakes and the extensive cracks that can spread across them. If you've ever had the experience of standing on a frozen lake when the ice fractures, you will know that the sound is otherworldly and awe-inspiring with the energy it injects into the environment. (If you have not had the opportunity to experience this, take a moment and do an internet search for the sound of ice cracking on a lake.) I like to imagine that this is akin to what it sounded like when the forces fractured in the early universe.[13]

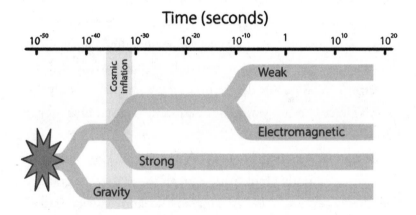

Combining the forces (at least three of them) together in GUT theories paints a utopic picture of symmetry, but there is trouble in paradise. We really like nice juicy predictions in the scientific method, and GUT theories make these aplenty. The problem is that the good folks who are doing the arduous work of testing these predictions keep getting negative results. For example, there is a prediction that—all by themselves—protons should decay into lighter subatomic particles. And many a PhD student in physics has labored through mountains of data and analysis looking for signs of this, but so far physics has come up empty-handed. At the moment, the best we can do based on the data is say that *if* protons can decay all by themselves, the half-life for this process is more than 10^{34} years. Just to point out the obvious, that is a lot of years—like 10^{24} times longer than the current age of the universe.

This is the landscape in which physicists are trying to understand the origin of the fundamental forces, and there is a lot of bushwhacking. The prevailing (though troubled) theory of "supersymmetry" (abbreviated as "SUSY," which is not too bad as far as acronyms go) dominates much of physics. Never underachievers, physicists have not just one but dozens of variations of SUSY models. The basic idea behind these supersymmetry models is that in the family tree of particles, every "normal" particle has a corresponding "super"-particle (for example, the electron would have a superparticle called the "selectron"). SUSY could do a significant amount of the heavy lifting to unify forces, create symmetries, and solve long-standing mysteries like what the heck dark matter is.

Then there is poor gravity, always the odd duck. As if the strong force weren't already problematic enough to get into the grand unified theory, we aren't even trying to shoehorn gravity into that scheme. Rather, unifying gravity requires a "theory of everything" (or TOE—I'm not kidding), which gets an A+ for hyperbolic names that can be misinterpreted by the uninitiated. In the figure on the previous page, you can see that *if* gravity was ever unified with the other forces, it split off basically in the Planck era—essentially at the same time the universe came into existence.

The upshot of this section is that we really don't even have much of a grasp of the forces as they are manifest in our universe, let alone what the actual underlying laws of nature might be, so we have a fair bit of work to do. This situation may remind you of the Game of Life, and the challenge of being a grid-cell scientist in that universe. It turns out that our universe may, in fact, be a simulation, in which case our predicament is not so different from the Game of Life at all.

Could the Universe Be a Simulation?

You may be experiencing significant skepticism toward the idea that a computer simulation could possibly generate the level of complexity we observe around us, which is absolutely a fair point. I don't entirely disagree, but you might correctly assume that there are some counter-arguments to consider.

Take humans, for example. I've found that when discussing whether the universe could be a simulation, people often immediately default to the assertion that "humans are way too complex to simulate," as if humans are the apex of complexity in the universe. We just can't seem to kick this nasty habit of assuming everything revolves around us. Even Descartes was already thinking about this in the 1600s. In *Discourse on Method and Meditations on First Philosophy*, he writes:

> [H]ow many different automata or moving machines could be made by the industry of man . . . For we can easily understand a machine's being constituted so that it can utter words, and even emit some responses to action on it of a corporeal kind, which brings about a change in its organs; for instance, if touched in a particular part it may ask what we wish to say to it; if in another part it may exclaim that it is being hurt, and so on. But it never happens that it arranges its speech in various ways, in order to reply appropriately to everything that may be said in its presence, as even the lowest type of man can do.[14]

To be sure, humans are complex. At least compared to something like microbes. The human brain is estimated to have roughly 86 billion nerve cells,[15] and with the number of connections each neuron has, the complexity of the human brain is hard to fathom. However, all the biological complexity of the human brain ultimately comes from the 23 chromosomes, with a total of roughly 3 billion base pairs that carry instructions for the body to carry out; the basic structure of the human brain is hardwired into these complex molecules. My friends, that is a physical manifestation of code that could (in principle) be programmed into a simulation, even if we are not close to being able to do so now (for which I am personally grateful); for better or worse, but the machine computing power we can engineer is far from reaching its limits.

For comparison, let's look to the Game of Life and its little square pixels that blink in and out of existence; it feels fairly straightforward to regard this two-dimensional universe as "not alive." However, if we consider the attributes of what it means to be "alive" (flash back to the chapter on ET life), defining the cell patterns in this grid as "not alive" starts to become slightly less obvious; they are composed of cells, use energy, can grow, can reproduce, respond to stimuli, can evolve, and can die. There may even be patterns in the Game of Life that are able to maintain homeostasis (which is sometimes part of the definition of "life").

To be clear, I am not arguing that the patterns in this game are "alive," but rather that this exceedingly simple game starts to call into question whether simulated life or a simulated universe is so far-fetched. In fact, despite how complex we might like to believe ourselves to be, our behavior and interactions are increasingly replicable.

When I first started teaching this topic in class, these things called "chatbots" were almost unheard of in the normal population, and my students would marvel at the conversations they could have with artificial intelligence programs. In particular, a go-to for my class was a chatbot called "iGod," with a tagline of "Repenting made easy." Today low-end chatbots seem to pop up on most commercial websites I visit, ready and waiting to answer (or try to) any questions I might have about which product or service is best for me. If your experience is like mine, most of these off-the-shelf chatbots do a mediocre job, at best, of mimicking an actual human—they primarily seem to be coded to search the company's site for information I could find myself if I took the time to search around. But don't let the performance of these second-rate chatbots fool you into thinking we can't simulate more realistic interactions. Now, at the time that I am writing this, ChatGPT has taken center stage, and is capable of remarkably sophisticated answers, which has become a major source of concern at universities where this program can be harnessed for less honorable students to avoid learning essential skills like reading, research, and writing.

The original framework to determine whether a computer could "think" was proposed by Alan Turing in 1950, which he termed the "Imitation Game" (today we call it the "Turing test").[16] Pared down, the point of this test is to see whether a computer program can trick a person into thinking the program is really a human. Computer programs have been passing this test for at least a decade, and it is now generally acknowledged that the Turing test wasn't sufficient to identifying true "sentient" artificial intelligence.

Since learning about the Turing test some decades ago, I've found that it comes front of mind for me far too often—especially when I'm casually interacting with people at a social gathering—we humans can be stunningly formulaic. The next time you engage in small talk with a loose acquaintance or stranger, I hope you will think about the

Turing test, too. A typical interaction you could have with a stranger might go along the lines of:

> You: "Hi, how are you today?"
> Stranger: "I'm good. And you?"
> You: "Good, thank you."

Those opening lines (or variations on them) are essentially prescribed. There are a limited number of socially acceptable ways to start a conversation with a stranger. At this point in the conversation with the stranger, if you have no interest in talking, you can just go about whatever your business was before the greeting—unless you are stuck in a situation through some social courtesy (like you are standing together at a cocktail party), in which it would be rude to just walk away. The socially acceptable topics you can bring up next are also highly constrained and generally limited to potentially shared experiences or commonalities—cue subjects like the weather, the local sports team, or the food at the party where you are both standing, which lead to another volley of dialogue:

> Stranger: "Did you get a chance to enjoy the nice weather today?"

Now there is a choice to be made. If you respond with something like "I did, it was lovely," you have offered no more information, and the social signal you are sending is that you are not really that interested in going down that path of conversation. On the other hand, if you respond with "I did, I had the chance to go on a great hike," you have opened the door for the conversation to continue; you have added information that builds the complexity of the dialogue ever so slightly. You have essentially invited the stranger to ask you about your hike, from which a set of new conversation topics could emerge. Now it is the stranger's turn to determine the conversation's path—they can ask you a question about the hike, such as "That's great, where did you go hiking?," or they can shut that path down with "That's great, I'm glad to hear it."

At each point in this small-talk conversation, there is essentially a formulaic flowchart unfolding, with each node typically leading to options that either terminate a line of discussion or take the conversation to a slightly more complex level, from which the number of potential conversation topics and connections expand. Eventually, you might find yourselves deep in conversation about how your first pet died. But if you had started the conversation in the initial greeting with the death of your pet, the situation would have likely been fairly awkward for the stranger, in other words:

You: "Hello, when I was six years old, my dog was hit by a car."

In which case, the stranger might be putting a lot of mental energy into a polite response and an escape route.

The point is that even human interactions (as complex as we think they are) have a very high degree of programmability. To be sure, social norms vary across cultures, but this too can be programmed. Another piece of complexity in human interactions is that we are both more illogical than we might like to admit, and we make mistakes and have flaws. The ways in which these characteristics manifest is subtle, but still largely algorithmic. For example, you can program how often and under what circumstances a person is most likely to say "um." So, the next time you are engaged in small talk, ask yourself what social algorithm you are following, and consider whether the person you are talking to is fundamentally behaving differently than a good computer algorithm. For these reasons, and more, I really dislike small talk, which makes me a disaster at parties.

Setting humans aside, and thinking a bit more broadly, one might infer that simulating an entire universe would be more challenging than merely simulating humans (which, after all, we have cause to believe are part of the universe). Even with the limited computing power currently on Earth and the short amount of time programmable computers have been at our disposal, we already have some pretty nifty simulations of big chunks of the universe; these simulations code in the laws of physics as we understand them, start with the initial conditions that we believe were present in the early universe, and let the programs run on today's supercomputers for a long time. Such simulations are able to generate swaths of universe down to the level of individual galaxies that are virtually indistinguishable from the actual observed universe.[17]

Granted, these simulations are far from achieving the detailed level of individual people, but I think there is also a case to be made that we are in early days of simulations, and I'm not brave enough to rule out the possibility of this happening in, say, the next couple hundred years (or less). More importantly, our hubris is sneaking back in again—just because we humans haven't figured out how to do something doesn't mean it can't (or won't) be done—especially if we are considering some kind of superintelligence outside of our universe that did the "software" development providing instructions to our universe. I would like to think that, if such an intelligence exists, they are more clever than we are. In this scenario, a designer of the universe could well be the equivalent of a precocious high school student in an after-school programming club, which begs us to consider where technological advancement might take us. Imagine what our ancestors of only a few hundred years ago would have thought of the computers of today. Now imagine what we might think of computers of a few hundred years hence. The upper boundary of computing power in our universe is only limited by the laws of physics and the size of the universe.

Today we tend to equate "computers" with circuit boards and hard drives, but this need not be so. In fact, the original "computers" were *people* who did calculations—famously, in the late 1800s the Harvard Observatory had a bank of women known as the "Harvard Computers," among whom were the famed pioneering women astronomers Annie Jump Cannon and Henrietta Leavitt.[18] Ultimately, a computer can be defined as anything that takes input, applies an algorithm, and produces output. This could be a MacBook Pro or a human. It could also be the solar system or the universe. Even the simple Game of Life is "Turing complete," meaning it can astoundingly carry out any algorithm.

We have many fancy programming languages to interact with modern computers, but at their cores, even today's computers use "assembly language," which is fundamentally binary—1s and 0s. The cell is on or off. You may not know what is being computed, but that doesn't mean that it isn't computing. In all these cases—human, laptop, solar system, Game of Life—an intermediary is taking input and generating output based on rules or laws. Taken to its conclusion, this line of thinking leads to the idea of "pancomputationalism" (of which there are many variations)—the idea that anything and everything could be thought of as a computer, taking input and producing output. Venturing into the realm of the highly speculative and unsettling, the entire point of our universe could be to do a complex calculation.

You'll be relieved to know that there are scholars who actually study topics like whether the universe could be a computer. One particularly staggering idea that came out of these studies is an entire philosophical argument (the "simulation hypothesis") that considers the probability that we are in a simulation.[19] This argument is highly debated, because—among other reasons—it appears to be untestable (as a reminder from the chapter on epistemology, not being testable should not be conflated with being wrong), at least unless there is a repeatable programming error that could manifest something entirely bizarre that we can empirically test. I don't know about you, but if our universe is a computer simulation, I really hope there are no bugs in the software.

The Simulation Argument is fairly nuanced and I won't reproduce it here, but the logic results in one of the following three things almost certainly being true: (1) we will go extinct before we harness the potential computing power of the universe, (2) our descendants living in an advanced technological state will have no interest in and or ability to run simulations like our universe, or (3) we are almost certainly living in a simulation. I am not a fan of option one, but I also don't think it is unlikely given our current trajectory. If I extrapolate human behavior today, option two seems extraordinarily unlikely by *choice*, but could result from us just not having sufficient resources to carry out many (if any) simulations at this scale. But, if we discount the first two options, option three becomes statistically likely. Why? Because universes simulated with this unfathomably advanced technology would most likely be easier to produce and greater in number than actual "real" universes.

To be sure, there is pushback on this argument, with criticism on various fronts, with some scientists even calling it "pseudoscience"[20]— though I am impelled to make a pedantic note that "pseudoscience" is not the same as "not being scientific"; the former implies that a position pretends to be scientific, the latter merely that the hypothesis isn't testable. There are also a range of more philosophical objections, including the identification of an apparent contradiction along the lines of "If humans are typical and we haven't created simulated universes, these types of simulations must be rare." One notable problem is that this argument only holds if we present-day humans have reached the apex of our technological advancement, which I highly doubt.

Even if we take the position that our universe is a simulation, we still haven't homed in on the laws of nature. *If* our universe is a simulation, and the physical laws that manifest for us come from the "software" that tells the universe how to behave, all we have really done is kick the can down the road. We might be able to say why our universe has the empirical laws that it does, but we are left wondering what the laws are in the metarealm where the simulation was built and where *those* laws came from. So, the laws of nature are still at least one turtle down.

The Trouble with Being in Our Universe

One of the unavoidable problems with trying to understand the laws of nature in our universe is that we are part of the universe, which gives rise to a number of potential philosophical entanglements. Logic (and language) are rife with paradoxes and problems that require a "meta" logic (or language)—*outside* of the logical (or linguistic) system—to detangle. As you will see, we end up in a bizarre situation in which we can *know* statements to be true but be unable to *prove* them. So much for the justified-true-believed framework from the chapter on epistemology.

As an example of the insidious problems that can pop up in a logical (or linguistic) system, consider the following statement:

"This sentence is false."

If we break down the above sentence into formal logic, we end up with:

A = "not A."

Which brings us back to Aristotle's laws of thought and the law of noncontradiction—a cat cannot both *be a cat* and *not be a cat* according to this logical structure. We already had a brief encounter with the potential limits of logic in the chapter on epistemology. The crux of the issue with sentences like "This sentence is false" (generally referred to as the "liar's paradox") is that when a subject refers to itself, we can get stuck in a paradox of self-reference.[21] A number of potential resolutions to the liar's paradox have been proposed, including "fuzzy logic," in which truthfulness of sentences like the one above is not simply binary (true or not true), but rather can have nonbinary values (i.e., half-true and half-false).

The mathematician and logician Alfred Tarski reached the conclusion that the liar's paradox can only happen in semantically closed language systems—meaning all statements or propositions within the language cannot be evaluated and understood solely within the system itself.[22] In other words, a language cannot define its own truth predicates. Instead, Tarski proposed that the liar's paradox can be avoided if there is a *metalanguage* that hosts the predicates for "true" and "false." You need to get *outside* of a language to assess the truth of statements within it. Sound familiar?

A notable issue with Tarski's method to avoid the liar's paradox is that it turns out the liar in this paradox has a cousin who is a barber:

"The barber shaves all of those, and only those, who do not shave themselves."

Does the barber shave themself? This is a version of "Russell's paradox," which involves mathematical sets. The paradox goes like this: Let's imagine that there is a set called "R" that contains all sets that do not contain themselves. Does R contain itself? If R does not contain itself, then it fulfills the condition of being a set that does not contain itself, which means it should be part of R. Or if we are talking about a hierarchy of languages à la Tarski, can there be a metalanguage that contains the truth predicates for all the languages that don't contain their own truth predicates?

I realize that it might sound like we are talking about semantics and language (and we are), but really we are also talking about the universe, which we guileless humans are trying to understand through the lens of logic and language (both symbolic and natural). Tarski's resolution for trying to understand a system we are part of might work on paper, but in the short term would require us to access a metaverse, and even in this case we ultimately just vest the issue with the next turtle in the stack. In the meantime, Russell's paradox suggests that we can't have a set of all sets—or in the case of the universe a law of all laws. Dealing with infinite regression is like playing whack-a-mole.

There turn out to be Gordian knots in symbolic language that can bring our entire understanding of logic and math to the breaking point. For example, consider this statement:

"This statement cannot be proved."

In the early 1930s, Gödel's incompleteness theorems took center stage and brought logic and mathematics to their knees. Aside from being shocked to learn that logic and math have knees, you may also be amazed that it was possible to subdue them with a single sentence. Despite the brevity of these two theorems, they have a density akin to solid lead. Here they are verbatim:[23]

First incompleteness theorem. "Any consistent formal system F within which a certain amount of elementary arithmetic can be carried out is incomplete; that is, there are statements of the language of F that can neither be proved nor disproved in F."

Second incompleteness theorem. "For any consistent system F within which a certain amount of elementary arithmetic can be carried out, the consistency of F cannot be proved in F itself."

At the risk of some minor imprecision, but in slightly more accessible language, what these theorems tell us is that in any logically self-consistent system, there will be basic truths that are not provable and that there are statements that cannot be proven within the system itself. In other words, there will always be facts that cannot be proven. For example, take the statement above, "This statement cannot be proved," and think about its incredible strangeness. At first glance, it seems to just be another version of the liar's paradox, but this statement is even more devious—it is true that you can't prove this statement, because if you did, the statement would be false. Which means the statement is true. *You can know the statement to be true without being able to prove it to be true.* Thus, in the scheme that Gödel revealed, there are statements that are unprovably true. In the context of epistemology, we now have a problem with the justified-true-believed framework if it is impossible for some particular true thing to be logically justified.

These incompleteness theorems meant that mathematics could not be completely formalized and reduced to a set of axioms and rules, which raised deep questions about the nature of mathematical truth, provability, and the foundations of logic. So, the next time you are in a math class and feeling snarky, go ahead and remind the teacher that mathematics is fundamentally incomplete and quote Gödel.

If the implications of Gödel's incompleteness theorem are starting to smell a bit ontological to you, you are not alone. These theorems actually caused Stephen Hawking to change his mind about an ultimate theory of everything, concluding that because our physical models of the universe are rooted in mathematical understanding, our physical understanding must also remain incomplete.[24] Nobel laureate Roger Penrose (who was born the same year the first incompleteness theorem was published) argues that "there must be something in the physical actions of the world that itself transcends computation."[25] Penrose and others have invoked Gödel's theorems to argue that the human mind cannot be mechanistic.[26] To be fair, these arguments have been heavily debated, but the point is that there is something perplexing going on that has had great thinkers thinking.[27]

Still here we are, mired in this apparently four-dimensional universe trying to understand the laws of the universe we are a part of. We only have access to how these laws manifest themselves within a limited range of physical conditions, and we are trapped within the system itself, necessarily leading to a problem with self-reference. Not only do we not understand the laws of nature now, but we may not even be capable of such understanding, and thanks to Gödel, that understanding could never be complete anyway.

Even if we did know what the laws of nature are, we are still left with abstruse quandaries such as determining where such laws reside and how they actually inform the action of the universe. Then there is the infinite regression that we just can't get away from: Are there laws that determine the laws? Could these laws have been different?

Interstition

Throughout this book we have continuously encountered properties of the universe (and its constituents) that have particular behaviors and characteristics—from the strength of the forces to the dimensionality of space-time, to the accelerating expansion of space. In each one of these instances, we might pause to consider what the universe might be like if these behaviors and characteristics were tweaked. How would things have played out differently if gravity were not this odd duck but instead had a strength similar to the other three known forces? What if we had an additional spatial dimension? Or the era of dark energy had started much earlier in time?

On one level, these questions might just appear to be entertaining thought experiments that make interesting homework assignments for my students. But, as always, there is more to the story; when we put the characteristics of the universe under a microscope, we find that changing them—sometimes just a tiny amount—would make life-as-we-know-it impossible. This leaves us asking why the properties of the universe are what they are, whether they could have been otherwise, and if the universe itself is fine-tuned.

10

Is the Universe
Fine-Tuned?

If you crack open an introductory physics book, you might find pages with many tables full of so-called "universal" constants, often known to high precision.[1] Of these constants, the ones you are most likely to have encountered without going into advanced physics are the gravitational constant ($G = 6.67408 \times 10^{-11}$ N m^2/kg^2) and the speed of light ($c = 299{,}792{,}458$ m/s), which are pretty essential in physics, even at introductory levels. I suspect that a typical physics student thinks something along the lines of "Some scientists somewhere figured these numbers out" and leaves it at that. But have you ever wondered *why* these parameters have the values that they do? Why these particular values (sometimes with many significant digits) and not something a little different?

The answer is *we don't know*, and that really bugs us. It is particularly troublesome because if you tweak any of these, sometimes by even just a little bit, it would profoundly change the properties of the universe, and in almost all cases make the universe inhospitable to life-as-we-know-it. These physical constants provide deep fodder for thought experiments that ask, "What if these fundamental parameters we take for granted were different?"

If you have ever played a stringed instrument, like the violin, you probably appreciate the importance of "fine-tuning." These tiny little metal tuners at the end of the string allow you to be able to tune the instrument *just right* so that it isn't discordant. If you "fine-tune" something, that means you make very small adjustments to get it to work most effectively. Given that many of these physical constants have to be *just right* for life-as-we-know-it, the question arises as to whether the universe itself is fine-tuned, which clearly has philosophical and metaphysical implications.

What Is Really Fundamental?

Physicists will be quick to point out that many of the constants we use in physics are actually related and can be derived from each other. For example, the speed of light (c) and the "vacuum magnetic permeability" (μ_0) had a child that they named "vacuum electric permittivity" (ε_0), which does kind of suck as far as names go, but I've heard worse. And the "fine-structure constant" (α) is the sibling of ε_0, h, c, and e. You might also think of the masses of individual particles as being fundamental, like the electron and proton, but their masses are more fundamentally dependent on other constants. In other words, mass is not actually fundamental, but depends on other properties of the universe. Think about that the next time you step on the scale.

Even though many of the constants you might find in your physics book[2] are not actually thought to be *fundamental*, the nice people of physics wanted to make things easier for you by including all of them in a one-stop, easy-to-look-up place. However, even when we've accounted for mutual codependencies, there are still about twenty or so independent (we think) constants that govern the universe (the exact number depends a bit on what is considered fundamental, so let's just say for now it is around twenty, which is plenty big enough).

The values of these twenty-ish constants get held up to a standard called "naturalness." The concept of naturalness is woven throughout physics and has unmistakable philosophical undertones and implications for the fundamental constants.[3] What naturalness refers to is an expectation that ratios of dimensionless parameters should "be of order unity." (In your talk-like-a-physicist lesson for the day: physicists say "of order unity" a lot. This is just an erudite way of saying "roughly equal to one.")[4] In fact, we've already encountered an echo of this when talking about the four known fundamental forces—three of the four forces are within a factor of 10^5 (which is admittedly not of order unity), and then gravity is way over there in the corner with a value that differs by a factor of more than 10^{30}. This dramatic difference intuitively signals that something funky is going on. In other words, the relative value of gravity violates naturalness.

Taking the concept of naturalness one step further, it can also produce very small and very large numbers. For example, if two values are very similar (but not identical) and some process pits them against each other, they will mostly (but not completely) cancel out—leaving a very small number. If there is yet another physical process that takes the inverse of this very small number, we end up with a very large number. For parameters that have values that can be roughly attributed to one of these paths to "naturalness," physicists are less inclined to worry about fine-tuning. What are especially troublesome are values that can't cleanly be "naturally" related to other parameters.

You may also be wondering what constitutes "small" and "large" in these contexts. The short answer is that physicists lean heavily into the Planck scale—parameters that are significantly larger or smaller than the Planck time or Planck length scale need to be explained somehow.

At the end of the day, we are left with twenty-ish constants running around without justification for their values, which leads to some big questions: Why do these constants have the values they do? Could they have been different? What would happen if they were? How do we explain that the universe appears to be so perfectly tuned for life?

What If These Constants Had Different Values?

Let's imagine we can play God and tweak the values of these constants in a toy universe and see what happens. Because the universe is complicated, many of these constants are intertwined in different physical processes, which adds an extra challenge to thinking through the implications of tweaking them. For example, both gravity and the strong force are important in nuclear fusion. The most straightforward option is to imagine an alternate universe in which we only tweak a single parameter at a time, but there is a large multidimensional landscape to explore with some give-and-take between parameters. To this point, you can think of the examples that follow as just a small sampling of parameter space to illustrate how variations of the physical constants we know and love might play out.

Shall we start with stars? They seem pretty important to the universe as we know it. Thinking back to the chapter on ET life, we know that light from stars is a convenient way to both enable the existence of liquid water and provide energy at the very bottom of the food chain. We also need these stars to hang around for a while before they die, or life might not have enough time to get going and evolve. For stars to help us out with this, it is important that they are in a state called "hydrostatic equilibrium." This is just a fancy term that means that the gravity pulling in is balanced with the pressure from nuclear fusion heating the core pushing out. If a teacher tells you to memorize this term and you are not an astrophysics major, I am really sorry.[5]

If gravity were stronger, a star would burn through its fuel faster (or potentially go straight into a black hole if gravity were really dialed up). In this case, stars wouldn't live very long, and it would be challenging for life to get a foothold before its host star expired.

On the other hand, if gravity were weaker, it would be harder for stars to form to begin with. If a star did form, it would burn through its fuel more slowly, providing less light and energy, but living a lot longer. Well, the living longer part is good, right? The catch is that we need carbon to make these nice complex molecules that life-as-we-know-it requires. To have carbon, we need stars to have lived and died and spewed their guts around the universe.

With this hydrostatic equilibrium example, we can go through similar thought experiments if we dial up or down other parameters, like the strong force or the strength of electromagnetic repulsion. In fact, if the strong force were too strong, all the hydrogen in the universe might fuse into helium before stars even had a chance to form.

It turns out that if we put the fundamental constants involved in star formation under a microscope, there is only a small range of parameters in which stars can exist at all.[6] However, because we don't know why these parameters have the values they do, we also don't know what the full range of *possible* values there could be.

The existence of carbon is also essential to life-as-we-know-it. If you've ever had the chance to hang out with science majors who are required to take a class called "Organic Chemistry," you have probably heard nightmares. The mere existence of carbon is to blame for much of this contempt. Carbon makes things like molecules, homework sets, and life exceedingly complex. As a reminder from the chapter on ET life, we need carbon if we want to have complex life (or really any life at all as far as we can tell).

However, the mere existence of carbon also appears to be an incredible coincidence. In fact, the coincidence is so acute that Fred Hoyle (of "Big Bang"–minting fame) declared it was a "put-up job."[7] Carbon is made in the cores of dying stars through what is called the "triple-alpha process."[8] In this process, first beryllium is formed from two helium nuclei (aka alpha particles), then another helium nucleus joins in the fun and the result is carbon:

$$\substack{4\\2}He + \substack{4\\2}He \rightarrow \substack{8\\4}Be$$

$$\substack{8\\4}Be + \substack{4\\2}He \rightarrow \substack{12\\6}C$$

That is all well and good, except for one catch: beryllium is *highly unstable*—it decays with a half-life of about 10^{-16} seconds (consider how insanely fast that is), and the likelihood that enough of the second reaction could take place during those 10^{-16} seconds is vanishingly low. The only way the triple-alpha process could work is if carbon just happened to have a resonance at an energy of exactly 7.7 MeV, which Hoyle predicted must exist (and is now named the "Hoyle state").[9] Subsequent work has found that the physical constants involved in this interaction must be within a range of 0.5 percent to 4 percent of the values they have if we want organic chemistry courses to keep being offered.[10]

Have you noticed how protons and neutrons have *almost exactly* the same mass? Many physics students have been grateful for this similarity, which makes solving some types of problems a tad easier. It isn't until the third significant digit that we encounter a difference:

Mass of a neutron = $1.6749274 \times 10^{-27}$ kg

Mass of a proton = $1.6726218 \times 10^{-27}$ kg

After being astounded that we can measure particle masses with this precision,[11] you might wonder why their masses are so similar, and what might happen if they were flipped and the proton were more massive than the neutron. Actually, the only difference between a proton and a neutron is the identity of a single quark (the neutron has two down quarks and one up quark, the proton has two up quarks and one down quark), so really this fine-tuning example comes down to the minor difference between the up and down quarks. If we switched their masses, we would alter the decay paths of protons and neutrons. In other words, in the universe as it is, neutrons can spontaneously decay into protons (plus other stuff) because protons have a lower mass.[12] In fact, when a neutron is stuck outside of a nucleus (where the strong force stabilizes it), it decays into a proton with a half-life of about fifteen minutes (i.e., much shorter than the age of the universe). While protons can capture an electron and turn into a neutron, free protons do not decay into neutrons spontaneously in the same way.[13]

If protons could spontaneously decay into neutrons, then we wouldn't have as many protons. So what, you ask? It isn't like protons are puppies. Well, for starters, if we didn't have protons, we wouldn't have puppies. Protons perform this essential task of allowing for the existence of atoms with well-behaved electrons whizzing around them. If we didn't have atoms, we wouldn't have molecules. If we didn't have molecules, we wouldn't have puppies (or any other life-as-we-know-it).

So that tiny little difference in the third significant digit of proton and neutron mass is important.

The dimensionality of our universe is also easy to take for granted, but if we were to change the number of dimensions of space or time, the resulting universe would not be so great to live in. If you naturally find yourself thinking about what the universe would be like if it had a different number of temporal and/or spatial dimensions than it has, then just plow right ahead (and also maybe think about getting an advanced degree in abstract math). For everyone else, thinking about what things would be like in versions of the universe that are almost impossible to imagine is a little more tricky, and you may want to glance back at the chapters on time and dimensions. The good news is that this is one of those topics where math comes to the rescue.

In the universe as it appears to be, we have a nice comfy set of dimensions. We have the three dimensions of space, and the single dimension of time. But why? When we talked about the theory of relativity in Chapter 6, one of the take-home points was that time and space are kind of the same thing. So why don't we have three dimensions of time and one of space? Or a nice matching pair of each? Or why not more dimensions altogether? As with the other examples in this chapter, we don't know.

Let's consider what would happen if our universe fell into a different box in the space-time graph below. As a thought experiment, what if the universe had fewer than three macroscopic spatial dimensions? Like, say, two. Here we run into issues with the structure of complex life. For life-as-we-know-it, there needs to be a way for things like nutrients to be distributed and waste removed (unless you're a face mite, which can live on your skin for about two weeks, camping out in your pores—but they don't have anuses, so they don't poop; they just store up the waste until they die).[14] This "requirement" for transporting nutrients and waste means there need to be pathways.

If we take a 2D cat and try to create simple pathways that distribute nutrients (or whatever) to every single part of the cat, we seem to either turn the cat into a ribbon (but a nice fuzzy ribbon!), or we divide it into lots of small pieces. There are some more creative options, too. For example, individual cells could have portals/valves that open and close to enable the transmission of material without the cat losing structural integrity. Perhaps you can think of some additional alternative models. Regardless, without the third spatial dimension to enable pathways that don't bisect the cat, it seems extremely challenging (but maybe not absolutely impossible) for a sophisticated feline to exist in 2D space. However, as a quick check on your hubris, if there are intelligent creatures living in four spatial dimensions, they might think similarly about us, and how "simple" 3D life would have to be.

So fewer dimensions may not be a good idea, but what about a universe with *more than three* macroscopic spatial dimensions? Here we run into other problems.[15] Take orbits, for example. You might just take for granted things like the Earth orbiting the Sun (at least I hope you know that the Earth orbits the Sun; if not, it's about time you learned this). The issue is that if we add extra spatial dimensions, even basic orbits are unstable (for fun, you might want to know that our solar system is actually unstable even in 3D; over many billions of years the planets will collide or be ejected), which is known as "Bertrand's theorem."

We already talked about how important a nice, long, stable life around a star is for life-as-we-know-it. This is not even remotely an option if orbits aren't stable. In the case of orbits in more than three spatial dimensions, our hapless planet would most likely get flung around the universe with abandon. It might occasionally be briefly grabbed by a star, only to be sent like a slingshot back out into deep space for who knows how long before encountering another star with which to repeat the process. The alternative is that the planet doesn't get flung out, but pulled in, where it is subsumed by the star. Those are the options. Could life evolve and survive under those conditions? Maybe. But probably not life-as-we-know-it.

If we don't think either more or fewer spatial dimensions will work out well, we are only left to play around with the dimension of time. The universe as we understand it has one dimension of time. One is a nice number—it's easy to work with, multiply, divide, you name it. Just to be abundantly clear, the alternative numbers of possible temporal dimensions besides one are zero or more than one.

Zero is the easy case, so let's start there. If there is no dimension of time, there can be no before or after, no causality—*nothing can ever happen*. That seems like a not very interesting universe to live in, and I am going to go ahead and conjecture that it would be hard for complex life to thrive in that environment.

More than one dimension of time is harder to think about. Intuitively, you might already be thinking that things would be a lot crazier if at any point in time you could turn temporally left. You would be correct. Here math comes in to save us again. If you haven't had a course in partial differential equations yet, you are in for a treat. Literally—go get a snack, and then come back.

Partial differential equations (or PDEs for those who are close friends) enable us to think about what would happen if we were to have different numbers of space-like and time-like dimensions. These types of equations have four main classifications that are called "elliptic," "parabolic," "hyperbolic," and "ultrahyperbolic." Which category a particular equation falls into in the case of dimensionality of the universe depends on how many dimensions of space and time it has.

In the universe we live in, we have three parameters that are space-like and one that is time-like: Space1, Space2, Space3, and Time1. This version of a PDE, with all but one of the parameters having the same sign, is called "hyperbolic." Hyperbolic PDEs are useful because they allow propagation of things, like light, or information. As far as I am concerned, this is the Goldilocks of PDEs.

Another interesting concept arises from a set of space and time dimensions that are hyperbolic; an alternate universe that had *three time-like* dimensions and *only one space-like* dimension might be OK. In this case you might try to think about this intuitively as time behaving like space as we know it, and space behaving like time as we know it.

If, on the other hand, we had *only* space-like (or only time-like) dimensions—Space1, Space2, Space3, Space4, and so forth—the equation would be classified as *elliptic*. Pretty much the only thing these types of PDEs are good for is describing equilibrium states, which makes sense, because without a parameter in there for something like time, nothing is going to happen.

The rubber hits the road when we allow for *more than a single parameter* of the same type. For example, what if our universe had Space1, Space2, and Time1, Time2? We find ourselves dealing with what are called "ultrahyperbolic equations," and these are pathological. In general, they do not have well-defined stable solutions. Another way to say this is that nothing would be predictable. The resulting chaos may not be too far off from what you envisioned intuitively when you thought about being able to turn left in time. Once again, we are faced with asking whether life could evolve under such circumstances, and once again, the answer is probably not life-as-we-know-it.

We have now effectively ruled out all the pixels in the space-time graph except for two. If we want a universe that can host life-as-we-know-it (which I am in favor of), then we either need three macroscopic dimensions of space and one of time, or three of time and one of space.

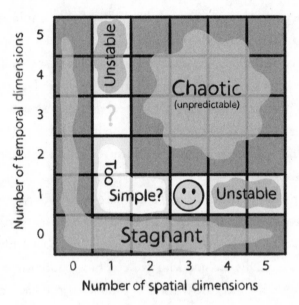

Chart inspired by Tegmark. Credit: M. Tegmark, "On the Dimensionality of Spacetime," *Classical and Quantum Gravity* 14, no. 4 (1997).

Not all the parameters in the universe that appear fine-tuned are specific values like masses of particles or numbers of dimensions. A prime example of this is a characteristic of the universe just after the Big Bang that has had an enormous impact on its habitability today—the primordial density fluctuations.

Even the very early universe couldn't escape from quantum mechanics and the reality that nothing can be precisely located in space and time, which meant no fields could be *precisely* flat. Instead, because the laws of physics said so, quantum fluctuations were imprinted on the mass-energy of the universe from the very first moment. These exceedingly minor fluctuations then proceeded to self-amplify with a positive feedback loop; little patches with more mass-energy had correspondingly more gravity and were better able to attract more mass-energy. When they attracted more mass-energy, they had even more gravity, and a snowball effect ensued. Apparently "the rich get richer" has been a theme in the universe since the Big Bang. Then the cosmological inflation kicks in, and all of these "small-scale" perturbations get dramatically stretched out to macroscopic scales.

In a classic coupling between the smallest and largest scales, the amplitude of these teeny tiny initial quantum perturbations determined the fate of structures that have formed over the course of cosmic history. In the recent literature these perturbations are commonly referred to with the parameter σ_8.[16] The value of σ_8 can be measured using observations of the cosmic microwave background, from which we have a current best value of $\sigma_8 = 0.82 \pm 0.02$.[17]

If σ_8 had been significantly smaller, the universe would be largely devoid of structure (which includes things like our own galaxy); the formation of stars would have been significantly inhibited; and consequently very few, if any, heavy elements would be available.[18] On the other hand, if σ_8 had been significantly larger, massive structures would have coalesced early in the universe, and much of the matter in the universe would have quickly collapsed into black holes. For any galaxies that did manage to form (and survive), they would have such high stellar densities that planetary systems would likely be unstable—any passing star that got a little too close would disrupt orbits and send planets off into the galaxy.

Let's not forget the very small anti-matter–matter asymmetry that came up in Chapter 7, which gifted us with something like 1/1,000,000,000 of normal matter particles surviving annihilation.

This one-in-a-billion difference is known as the "baryon asymmetry" (typically denoted with the symbol η_B). A "natural" number for this asymmetry would have been 0, but if it isn't 0 it could (in principle) be anything between 0 and 1.

We've already established that if the baryon asymmetry were exactly 0, you wouldn't be reading this book and there wouldn't be paper for the book to be written upon. But what if this asymmetry were even larger? At the most basic level, if we crank up η_B, less matter is annihilated (and turned into light), which has a host of downstream effects, starting with primordial nucleosynthesis and structure formation.

Having a higher density of elementary particles available for nucleosynthesis enables more reactions to happen in those first few moments after the Big Bang, which results in a lower relative abundance of hydrogen and higher abundance of helium (there is a classic paper on this from Robert Wagoner in 1973 that is full of interesting tidbits on what happens).[19]

Why do we care if there is less hydrogen and more helium?[20] If the abundance of hydrogen is significantly reduced, then nuclear fusion in stars gets much harder, and consequently the duration of time that stars undergo nuclear fusion is decreased. As we saw in the chapter on ET life, longish stellar lifetimes may be important for life to have time to emerge, so if we dial up the helium too much we are in trouble.

Increasing the mass-to-light ratio also impacts the structures that can form in the universe, which also places limits on potential habitability. For example, if η_B were too high, stellar densities would be much higher, and as a result the likelihood of supernova exploding in the vicinity and destroying life (and everything else in its path) would go up. Recent work places an upper boundary on values of η_B that could be tolerated of 10^{-4}.[21] On the other hand, if η_B were too low, there would not be enough self-gravitation for stars to fragment out of galactic disks, which gives us a lower limit of 10^{-11}.[22] So based on our current models, we have wiggle room in the range of $10^{-11} < \eta_B < 10^{-4}$. Whether we consider this range small or large depends on the full range that might have been possible (which is possibly $0 < \eta_B < 1$).

Before we move on from examples of possible fine-tuning, I want to dwell on an exquisite property of our universe that I don't think gets nearly enough attention, but is critically important for life-as-we-know-it. Specifically, at the physical and temporal scales we live, our universe appears to be in a delicate balance between order and chaos. How complex patterns (e.g., life) arise from simple rules is a profound question (and takes us back to the Game of Life in Chapter 9).

To put a finer point on this, I want to borrow from work done in the field of information theory, and, in particular, blatantly usurp two terms that are central to this topic: "algorithmic compressibility" and "logical depth."[23] In the most basic terms, you can think of algorithmic compressibility as the extent to which an object has redundant patterns that could be made more concise. For example, a checkerboard pattern is highly algorithmically compressible; if you were to write a computer program to reproduce the board, it would only take a few lines of code. On the other end of this spectrum is logical depth. If you have ever accidentally deleted a term paper that you spent days writing, you have a visceral understanding of logical depth—the paper you lost will not be easy to reproduce, and a computer program would require an extensive amount of coding to reproduce it.

Our universe appears to be a delicate balance between algorithmic compressibility and logical depth. To see this, all you need to do is look at the nearest tree—symmetries and patterns are embedded in the leaves, bark, wood, and branches. However, at the same time, the patterns are not perfect, no two leaves are the same, and there are random deviations at every scale.

Of the many astounding facts about the universe, to me the balance of the two extremes of algorithmic compressibility and logical depth might be the most simultaneously profound and overlooked. The universe need not be so. If the universe were entirely algorithmically compressible, everything would follow perfect patterns and symmetries, and nothing interesting (a category in which I include life-as-we-know-it) would ever happen. On the other hand, if the universe were dominated by logical depth, there would be little (if any) law and order to reality, with the universe bordering on chaos. Yet, in the universe in which we live, we seem to have some tidy physical laws that govern reality hand in hand with chaotic behavior and random fluctuations imprinted by quantum mechanics.

Turning to information theory as an analogy, the most economical computing programs are the shortest necessary for the required outcome, and hence, the most economical programs are most likely to have resulted from a universal computer.[24] In a similar way, this philosophy sits at the root of physicists seeking symmetry and unified laws. Building a system that is both simultaneously deterministic and random requires a careful design—there needs to be enough randomness injected to break perfect symmetries, but not so much that it devolves into chaos. To be fair, this analogy starts to invoke a bit of teleology—that a thing exists for a purpose, with which I am not entirely comfortable, but I also don't deny the possibility that our universe is essentially a computer in the literal sense—taking input, running a program, and producing output.

I conjecture the apparent balance between algorithmic compressibility and logical depth is a requirement for life-as-we-know-it. If we were observing the universe at a significantly different time or spatial scale, we might not infer the same balance. For example, the physics of the early universe after inflation was very simple and uniform. Likewise, in the distant future when dark energy has taken over, the observable universe will also be simple and uniform. Or on physical scales approaching the Planck length, it might be hard to infer any rhyme or reason to anything that happened. Similarly, when modeling entire galaxies, there are large-scale patterns that dominate behavior. I'm left to tenuously conclude that there is something very special about the intermediate parameter regime we inhabit in the universe.

Given all these parameters that appear to require very specific values for us to be here, one could understandably feel that the universe must surely be a set-up job. However, before we can jump to that conclusion, we ought to take stock of lessons we have learned over the course of humanity that should have bearing on our thinking here. You know that saying about people who forget history being doomed to repeat it? This is that part of the text. Before we get to possible explanations for the apparent fine-tuning, it is important to take a step back and think about other times and situations that have played out for humans and the lessons learned.

Lesson 1: The solar system. The solar system itself was quite a source of consternation not too long ago. We have all these planets going around the Sun in nice "perfect" circles[25] in very nearly the same plane of the sky. These perfect orbits can't possibly have happened randomly! It's too ordered! Too perfect! As Newton wrote:

For while comets move in very eccentric orbs in all manner of positions, blind fate could never make all the planets

move one and the same way in orbs concentric, some incon-
siderable irregularities excepted which may have arisen from
the mutual actions of comets and planets on one another,
and which will be apt to increase, till this system wants a
reformation.[26]

In other words, the brilliant Newton thought the solar system
was a set-up job. But surprise! At that time, we didn't know all the
physics that go into forming a solar system. In model solar systems
that we make today using advanced (well, advanced to us today) computational techniques and all the physics that we currently know, it is
virtually impossible to make a solar system that does not have properties broadly similar to our own, with planets going around on nice,
near-perfect circles in the same direction and in very nearly the same
plane. It turns out that blind fate, once in the hands of physics, had
little choice but to "make all the planets move one and the same way
in orbs concentric." While the geometry and dynamics of the solar
system might have seemed fine-tuned a couple hundred years ago,
today we understand that physics made it so. Lesson learned: Don't
assume we humans know all the physics involved.

To be clear, just because we now understand the physics that naturally led to solar systems like ours, that doesn't mean that the broader
universe and the laws that govern it are *not* fine-tuned. Rather, that
the creation of the solar system out of the maelstrom of the universe
did not require the existing rules to be tweaked or bent for our system of nice concentric orbits to come about. More generally, we need
to be mindful of the logic we talked about way back in Chapter 2 and
denying the antecedent.

"If the solar system is fine-tuned then the universe is
fine-tuned."

If P then Q. Valid.

"If the solar system is not fine-tuned then the universe is not fine-tuned."

Not P therefore not Q. Not valid.

Lesson 2: The habitability of Earth. Earth is a pretty darn good planet as far as life-as-we-know-it is concerned. With fewer planets in the solar system than you can count on two hands, what are the chances that there would be such a downright perfect planet for us to live on? In every respect, the Earth seemed pretty darned special. Only a couple decades ago, we didn't even know for sure if there were other planetary systems out there. The first confirmed detection of an extrasolar planet (commonly shortened to "exoplanet") was in 1992, which demonstrated the existence of planets orbiting a pulsar (just FYI, that would *not* be a good place for life-as-we-know-it).

In the intervening decades, the search for planets in other star systems has exploded, and there are now literally thousands of confirmed exoplanets orbiting other stars in our galaxy. In the observations obtained to date, over half of these exoplanets are part of multiple planet systems, and roughly one in five Sun-like stars has a nearly Earth-sized planet orbiting at a distance that falls in the Habitable Zone.

All of that means that Earth is probably not really all that unique. When you consider the sheer number of Earth-like planets orbiting at Earth-like distances around Sun-like stars—just in our galaxy alone—our sense of importance gets knocked down a notch. Observations from the Kepler mission indicate that in the Milky Way, there are roughly 11 billion of these Earth analogs. If you could visit a different one of these Earth analogs every year, it would take you nearly the age of the universe to see them all.

Once you know that Earth is only one of billions of similar planets in the Milky Way, it no longer seems quite so special and fine-tuned for life. Lesson learned: Be wary of assuming yours is the only example and not part of a much larger ensemble.

Lesson 3: The human body. Even the human body provides lessons for us to consider, and they have long been used as evidence for fine-tuning. How could such complex systems have come about by random chance? A prime example is the human eye; our eyes are downright amazing. In many ways the human eye even dramatically surpasses astronomical detectors that we can make (at least for now). Consider the following:

- The eye has a huge field of view, nearly 180 degrees. With your peripheral vision you can see light from nearly the entire space in front of your head. In large part, this field of view is due to there being a curved "focal plane" and detector (aka the "retina") on the back of the eye that allows light coming in from different angles to still be in focus.
- Speaking of light detection, the human eye has approximately 125 million light-detection cells. You can think of these as akin to pixels in your camera. How many pixels does your camera have? Technology is advancing quickly, but at the time of this writing, the best cameras on the market have on the order of 50 million pixels and cost a pretty penny.
- With this many pixels, our eyes also have an astonishingly rapid readout, which is the time it takes for the

light the cells have detected to be registered by the brain. The cells in the human eye read out at roughly 30 times each second (or 30 hertz). By comparison, that 50 megapixel camera that is on the market now has a maximum readout rate of 5 frames per second.

- The range of faint to bright light the human eye can process is also astounding. This ratio of the brightest light detectable to the faintest light detectable is called the "dynamic range." The human eye has a dynamic range of roughly 1 to 10 billion, rivaling even astronomical detectors.

Given the capabilities of the human eye, it is hard to blame folks who see it as evidence of fine-tuning. The crux of the issue here is the power of natural selection to shape evolution over millions of years. Because we don't see significant changes over the course of a human lifetime, extrapolating from our own experience, it is hard to conceive of such amazing things as the human eye arising from evolution. Even Darwin has a quote in *The Origin of Species* that would seem to support this position:

To suppose that the eye with all its inimitable contrivances for adjusting the focus to different distances, for admitting different amounts of light, and for the correction of spherical and chromatic aberration, could have been formed by natural selection, seems, I freely confess, absurd in the highest degree.

But once again, be careful. That particular quote is frequently taken out of context. Compare the previous excerpt with the full quote from Chapter 4 of *The Origin of Species*:

To suppose that the eye with all its inimitable contrivances for adjusting the focus to different distances, for admitting different amounts of light, and for the correction of spherical and chromatic aberration, could have been formed by natural selection, seems, I freely confess, absurd in the highest degree. When it was first said that the sun stood still and the world turned round, the common sense of mankind declared the doctrine false; but the old saying of *Vox populi, vox Dei*, as every philosopher knows, cannot be trusted in science. Reason tells me, that if numerous gradations from a simple and imperfect eye to one complex and perfect can be shown to exist, each grade being useful to its possessor, as is certainly the case; if further, the eye ever varies and the variations be inherited, as is likewise certainly the case; and if such variations should be useful to any animal under changing conditions of life, then the difficulty of believing that a perfect and complex eye could be formed by natural selection, though insuperable by our imagination, should not be considered as subversive of the theory. How a nerve comes to be sensitive to light, hardly concerns us more than how life itself originated; but I may remark that, as some of the lowest organisms, in which nerves cannot be detected, are capable of perceiving light, it does not seem impossible that certain sensitive elements in their sarcode should become aggregated and developed into nerves, endowed with this special sensibility.[27]

What Darwin did or did not think about the human eye is a red herring in any case. Just because he was a clever guy who wrote an important book does not mean that his word is final. During Darwin's lifetime, we didn't even know about DNA. On the other hand, modern research has a much better grasp on the origin of eyes and can now trace the evolution of eyes back to light-sensitive cells in the brain that provided the host animal with an advantage for survival. Lesson learned: Don't underestimate the power of natural selection, even if you think you understand it.

However, we need to remind ourselves (as with each of the lessons learned above) that just because even amazing things like the human eye don't seem to require fine-tuning, that doesn't mean that the universe isn't rigged for our existence.

Possible Solutions to Fine-Tuning

There are a range of possible explanations for apparent fine-tuning. To be sure, I don't know which (if any) of these is correct, and I would argue that you don't, either. These options are also not exclusive of each other, so feel free to mix and match. To set the stage, we first need to discuss a concept called the "anthropic principle," which is at the foundation of the discussion of fine-tuning.

The anthropic principle was introduced by Brandon Carter in the 1970s.[28] In bare bones, the anthropic principle says that the properties of the universe in which we live must be compatible with us living in it. This smacks of a tautology, but at its core, it is an essential concept to keep in mind—known as "selection bias," generally, and "survivor bias" specifically.

For example, if you were to go out in a fishing boat and use nets to catch fish, if your nets have holes of, say, eight inches in diameter, you probably shouldn't expect to capture too many fish smaller than eight inches. Alternatively, after you haul in your catch, you could (wrongly) infer that there are only fish in the ocean that are larger than eight inches. Anything that affects the sample you end up with, causing it not to be a random sampling of the possibilities, introduces a selection bias. Like big fish.

Here's the thing—with selection bias in mind—we wouldn't have evolved and survived to observe the universe and ask whether it was fine-tuned if we didn't happen to be in a universe that allowed for us to survive and evolve. By analogy, we shouldn't be surprised to find only large fish in our net.

A version of this argument was presented in a famous paper by Robert Dicke in 1961, which you might not realize digs into fine-tuning based on the title, "Dirac's Cosmology and Mach's Principle." In this paper, Dicke works through a beautiful line of reasoning in an effort to understand why the universe is as big as it is, and he concludes that we would not be here if it were otherwise (in the following quote, "T" refers to the age of the universe): "Thus, contrary to our original supposition, T is not a 'random choice' from a wide range of possible choices, but is limited by the criteria for the existence of physicists."[29]

Dicke's reasoning goes like this: for us to be here asking why the universe is as big as it is, there must be carbon to make complex life. For there to be enough time in the universe for carbon to be created and distributed, the universe must be at least about 10^{10} years old. On the other hand, if the universe were significantly older, only stellar remnants would be left. Thus, there is only a narrow window in which we could be here asking the question of why the universe is as big as it is. Since light has a finite travel time, the observed size of the universe is necessarily the distance light could travel times the age of the universe. To be sure, his argument was rough, and we have learned more about the physics of the universe in the intervening years that have tweaked the precise values, but the underlying argument is still valid.

As presented here, this is a "weak" form of the anthropic principle. There are other stronger forms as well, which requires a refresher on Aristotle. Aristotle thought a lot about why things happen in the world. His thinking on this centered on the idea of *cause*, but he thought about cause in a much more nuanced sense than is common today. In fact, the idea of *cause* was so important to him that he came up with not just one, not just two, not just three, but you guessed it, *four* types of cause: "material," "formal," "efficient," and "final."

The material cause isn't too hard to get a handle on: it boils down to what something is made of. In the case of the universe, we might say that its material cause is mass-energy (which is then manifest in all types of fun and entertaining particles). Or in the case of using a net to catch fish, the material cause would be the net.

The formal cause is a little more abstract, but you can think of it as the rules or plan for something. The laws of physics might be, for example, the formal cause of the universe. These laws tell everything in the material cause (e.g., particles) where to go and what to do. Back to our fishing expedition—this would be the plans the net maker used to make the net.

The efficient cause is what or who makes it happen. This one starts to get tricky when we talk about the universe, clearly on the borderlands of science and in the territory of theology and philosophy. What or who breathed life into the universe to begin with? Why does any of it exist at all? And how do we get ourselves out of the infinite regression that follows? This is far more straightforward for our fish—the net maker made the net (or more likely a machine in a modern factory today, in which case an engineer made the machine that made the net).

The final cause is where we finally loop back to the strong anthropic principle. The final cause for something is its purpose for being. Why was it created at all? If we are talking about, say, a coffee mug, this isn't too hard to answer. When we are talking about the universe, as you surely appreciate, this is a bit more challenging. In a nutshell, the strong anthropic principle states that the point of the universe coming into being and evolving the way it has was for the purpose of sentient life emerging. This is a harder pill to swallow. I am not saying that the strong anthropic principle is wrong (who knows?), but it does have a hefty dollop of hubris embedded in it, which makes me skeptical. To the point, over the course of humans trying to understand the universe, every time we think we are the "center," we have been wrong. But this is far from an airtight argument against the strong anthropic principle.

One way or another, our universe seems to be exceptionally well suited to our being here, which is begging for an explanation. Here are some options for consideration.

Option 1: Chance. Pure random luck. Just because something is extremely unlikely, doesn't mean that it is impossible. Exactly how unlikely it would be to draw the current universe out of a hat depends on how many other things are in the hat. If only small ranges of each of the fundamental physical parameters are possible, then our particular universe may not be all that unlikely.

The real trick here is knowing (or at least constraining) the "possible range of values," which we generally don't know. Without knowing the range of possible values, we don't know how much real estate is available in parameter space, and consequently we don't know how special the exact value of these constants in our universe are.

Option 2: Providence. We can't talk about fine-tuning and not bring up providence of a higher power as an option. This need not be a conventional "god"—any higher-level supreme being or alien race will do. We just need some external agent to which we attribute some benevolence or desire to create a universe with life. This possibility is squarely in the realm of religion until it becomes testable. To be sure, fine-tuning arguments are frequently used to support the existence of a supernatural agent. This is actually just a thinly veiled "God of the gaps" fallacy. However, the fact is, we can't test it, which means we can't rule it out either.

Option 3: A simulated reality. Another possible explanation is an offshoot of providence, but I think a simulated reality merits its own paragraph. Imagine that you are a super-advanced being with unfathomable "computational" resources (whatever "computation" looks like in your super-advanced state of being). One day you're bored, because you've pretty much done everything there is to do in existence. You think, "I need to create something to entertain me! Maybe some pets who will amuse me for eons with their antics." I mean, you've seen the popularity of reality TV, right? So, you get on your multidimensional computer and start to build a toy universe. After trying several parameter values that were boring, or caused the universe to die, you finally hit on the right combination that would generate cute little pets. (In case you have forgotten, there is a whole section on the possibility of the universe being a simulation in Chapter 9, and this explanation is not so different from the Game of Life.)

Option 4: Unknown physics. The next possibility requires some humility on our part—acknowledging that there may well be unknown physics that has a role in determining the precise values of various parameters in the universe. Since we don't know why the fundamental parameters have the values that they do, we don't know whether they could have been different. The bulk of the discussion on whether the universe is fine-tuned assumes that these parameters were drawn from a large range of possible values. But we don't know this to be true. To be sure, we know we don't know a whole bunch of physics, so we need to hold space for more advanced theories being developed that could provide bona fide reasons for why things are the way they are.

I know that many physicists are holding out hope that at least some values for constants will fall out of string theory (or something like it), and then we can blame the Calabi-Yau manifold (or some shape like it) for the state of our universe. This may well happen. But . . . shifting responsibility to a particular Calabi-Yau manifold just kicks the can down the road, because the next question will be "Why that particular Calabi-Yau manifold?"

Option 5: Chauvinism. Speaking of humility, perhaps we are not nearly as important or special as we would like to think. One explanation for fine-tuning is that our imaginations are far too limited, and life *in some form* could evolve in universes with a vast range of values for the fundamental parameters. This is the reason I keep referring to "life-as-we-know-it," despite that annoyance of having to type that out instead of just writing "life." Sci-fi authors could be onto something when they talk about creatures like Crystalline Entities and even more bizarre life forms that might thrive in radically different universes. For example, consider the Boltzmann brains we encountered in Chapter 4, the underlying concept for which opens up a range of possibilities for life.

Option 6: Natural selection. What if universes with the fundamental parameters similar to those we observe in our universe are able to procreate and multiply like we saw in the chapters on black holes and other universes? And what if there is a type of heredity between a mother and child universe in which the child inherits similar values of the fundamental parameters? For example, if baby universes are able to form from the creation of black holes in a parent universe, this idea could result in a type of natural selection. In which case, universes with parameters that favor the creation of black holes would have a lot of offspring, while universes that have parameters that make forming black holes hard or impossible would be sterile. In this case, our universe might not be all that statistically unlikely.

Option 7: Multiple universes. Maybe universes exist "out there" with every possible set of physical parameters. If the set of possible physical parameters is large, this idea would suggest a very large number of universes. In this case, a universe must exist that has parameters that allow us to evolve. It shouldn't surprise us that we find ourselves in that particular universe because we sure as heck aren't going to find ourselves in a universe that makes life-as-we-know-it impossible. (Hat tip to the anthropic principle.)

Option 8: An ethical requirement. This is a teleological argument and goes straight back to the strong anthropic principle, relying on the universe having a "final cause" (à la Aristotle). In this case the final cause of the universe might be to create life like us. This option means the universe has a purpose, and perhaps that purpose is enabling our existence. That's a pretty nice ego boost, and it does feel nice to think we are the reason the entire universe exists. Unfortunately, just because we would like something to be true doesn't make it so. It could also be the universe exists to perform a calculation, or to perform an experiment, both of which are less ego-enhancing.

Option 9: Something else. Do you really think that our puny little human brains that typically only have seventy or so years to learn as much as they can, only exist in three spatial dimensions, and can't intuit quantum mechanics are actually capable of figuring out all of the possible options? Neither do I. Maybe you have already thought of additional options, or variations of those listed here.

Each of these options relies on different possible assumptions, some of which might seem crazier to you than others. In class, I will have my students rank the possibilities and try to justify their rankings, which gets us deep into the justified-true-believed framework and back to the epistemology we started this book with. For many people, the options that seem the least crazy are the ones they have had the most exposure to. For example, roughly 80 percent of adults in the United States believe in a higher power of some kind. For people in this category, the "providence" option probably seems fairly acceptable. Be careful, though: just because something is familiar to us doesn't make it a good explanation—this is a version of the availability error fallacy. A challenge is to ask yourself why some of these assumptions strike you as more or less acceptable and what your justification is for that position.

Interstition

We have journeyed through a series of abstract and mystifying topics. In grappling with the provocative possibilities of reality outside of our perception, we have come face-to-face with limits of science, logic, and our own intellectual capacity. The topics we've visited inexorably connect back to what it means to be human and our place in the universe; if this hasn't been a humbling experience for you, I have failed. However, this experience can also be vitalizing and even restorative if we let it.

How do we take our perspective on the universe, with all its philosophical tangles and persistent questions marks, and move forward?

11

What Is Our Place in the Universe?

Into the Woods

During recent years I have become obsessed with mushrooms. The sheer array of species that pop up over the course of a summer in my little corner of nature is wondrous, with colors, shapes, and sizes that put the standard gray-brown varieties of grocery stores to shame and rekindle a childlike desire to make fairy houses out of twigs and moss.

Walking the same trail daily, I witness the astounding rate of growth of these fungi blooms—and equally astounding rate of demise. The thing about fungi that I find most remarkable is that the mushrooms we see on the surface are only the fruit of sometimes vast mycelial networks that connect and interact with an entire ecosystem below the surface that is hidden from us. In fact, the honey fungus (*Armillaria solidipes*) is one of the largest known organisms on Earth.[1] One particular fungus mycelial network is known to span more than two thousand acres. Perhaps even more astonishing, it is believed to be more than two thousand years old.[2]

You are perhaps wondering why I am writing about mushrooms in a book about the cosmos. The answer is, in part, simply because I am fascinated by them—a big part of the way I keep my personal existential crisis in check is by the simple act of trying to notice the beauty around me. I know that I am treading precariously close to the content of a Hallmark card here, so let me dial up the nerdiness: one of my own mindfulness practices involves looking for math—patterns, symmetries, fractals, turbulence, harmonics, ratios, logical depth, or algorithmic compressibility—in nature. I will stop abruptly on a hike to inspect a flower that has an unexpected symmetry or trace the fractal patterns of a leaf, similar to the fractal patterns we see across all of nature—from coastlines, to star-forming regions, to the filaments of galaxies in the universe. Witnessing that the smallest and biggest things in the cosmos are connected and intertwined serves as a protective barrier against nihilism. Seeing the architecture of math—like the numbers pi and phi—manifest in nature over and over borders on a mystical experience. But there is more to my inclusion of mushrooms than merely their beauty and symmetries; they strike me as richly symbolic of our place in the universe.

I recently snapped a photo of a lovely little mushroom that had sprouted up through a small hole in a leaf and bloomed as if to celebrate its happy circumstance. A friend of mine captioned this photo "The Anthropic Principle," and my sense of the rightness of that caption was immediate. If any one of countless seemingly random occurrences in the universe had been arbitrarily different—even the softest eddy of wind—things would be otherwise.

The particular happenstance of that mushroom was compounded by how transitory it was. But even as the curious little mushroom had only a few brief days above ground, it is part of something much more expansive that permeates the hidden ecosystem below the surface.

I am left wondering whether we have a similar arc of existence in the universe—whether there is a web of life from which "intelligence" might spring if the conditions are right, whether we are part of something much larger and hidden from our limited perspective, whether we have a symbiotic relationship with the cosmos, and—if so—whether we are doing our part. Taking our happenstance for granted strikes me as tragically complacent. We are fundamentally collections of molecules imbued with something we call consciousness—fluctuations of universe that are self-aware. It seems a real pity to me if we don't take maximal advantage of our cosmic self-awareness during the brief time we have to bloom above the forest floor.

The question is, given that we have this cosmic awareness, what do we do with it while we are here? To be sure, one option is that we ignore it and go about our existence as if we are not part of something much more expansive than our paltry human experience might lead us to believe. I'm not a fan of a blanket ignore-the-cosmos approach, though, which you might have inferred from my career choice. I suppose the universe doesn't really care if we pay attention to it or not, but we ignore it at our peril. To begin with, in any matchup of humans versus the universe, the universe will *always* win. More importantly, given that we are sentient fluctuations of the universe, trying to understand the universe ultimately gives us insight into who we really are and what we could be. Ignoring the cosmos feels a lot like not living up to our potential.

The Fermi paradox is staring us in the face. If one of the options for solving this "paradox" is that technologically advanced civilizations frequently or even always destroy themselves, I would like to humbly advocate that humanity try not to be a data point in that sample. My belief is that, collectively, humanity needs more exposure to awe and wonder to draw our focus outward and remind us that we are connected to something unimaginably vast, that—much like the mycelial network—extends far beyond both our short lives and limited lived experience.

Unfortunately, and simultaneously, my sense is that our collective willingness to embrace existential questions of reality is becoming increasingly thin amid modern life as we spend less time in nature. I worry that this growing detachment from the reality of the universe and our place in it is slowly making us collectively myopic. Over the generations, we have increasingly moved inside at night, where we have a level of comfort and safety that can make us indifferent and incurious about the outside world. The night sky has all but disappeared for over 80 percent of the global population living under severe light pollution.[3] I can't help but wonder how this impacts our worldview and sense of self and place in the universe. I conjecture that this disconnection from the universe translates into an overinflation of superficial concerns. I could well be romanticizing the distant past, but I yearn for a time when people would look up at the sky at night and simply wonder.

I know that for many people, going about the business of being alive without considering the underlying reality of the cosmos is a comfortable condition. I can slip into that mindset, too, especially amid the urgent demands of life, so I'm not here to pass judgment. As I sit at my table writing, I am struck by the vertigo of simultaneously interacting with the mundane "reality" of the day-to-day, getting the laundry done, and being aware of the profound "true" reality that is outside of our experience and perhaps beyond our grasp. I suspect that refocusing on the day-to-day is in part a mental defense mechanism and a retreat to our comfort zone, and in part a purely practical necessity to continue with the tasks of daily life and our physical requirements as living creatures (after all, homeostasis only gets us so far, even for tardigrades). The rapid flow of daily life is always there to entrain us, refocusing our concerns on pressing daily issues from groceries to politics. But the pervasive tendency to ignore the reality of our place in the universe and the devaluing of the future of humanity can and will come back to bite us. Before we know it, the future is upon us. We can chastise our one-time selves (or our ancestors) for not having had a broader or longer vision, but that does us little good in the moment. Daily choices add up, and the integral of these choices across humanity not only determines our own path, but the circumstances our descendants will face. I am struck by how easy it is to discount our ethical responsibility to the people who will live in the distant future. Even just a few generations from now seems beyond the horizon of our concern.

Sometimes I feel that humanity is participating in a global-scale Marshmallow Test; in this classic psychology experiment, children are tempted with a marshmallow. If they can wait a little while and not eat the marshmallow, they will get a second marshmallow later. While this experiment is typically done with young children, it strikes me as comparable to the current evolutionary state of humanity and the choices we make that have impact on our flourishing in the long term. After all, we seem to be in our technological adolescence, and we haven't quite developed a societal prefrontal cortex. Our immediate aspirations continually outpace our long-term vision and ethical considerations.

This isn't really our fault—we are evolutionarily hardwired to be shortsighted; if you didn't eat that berry *now*, someone (or something) else would. Consequently, we are prone to heavily discounting the future. I chronically undervalue my time a month or a year from now, which I also chronically regret, yet somehow, I never manage to stop devaluing it. Immediate needs trump long-term what-ifs; it's hard to worry about the fate of the universe when you are trying to meet a 5:00 p.m. deadline or getting food for your family.

Intelligent Life?

When I think about the patterns and habits of modern life, a glass-walled ant farm my children once had often comes to mind. We would watch the ants scurry about, devoting their lives to digging tunnels and ferrying crumbs back and forth. As far as I could tell, the ants seemed perfectly happy with this state of being, but for all I know some among them may have been in existential crisis, questioning their significance and asking whether there was more to life than digging and collecting crumbs. Watching the ants dig new tunnels, some of which would abut the glass, I couldn't help but wonder whether any of them wanted to know what was beyond this invisible barrier and if they hoped one of their tunnels would get them access to something more. Or alternately, whether they just accepted the barrier they didn't understand and went about their business without a care for reality beyond their ability to access it. I liked to imagine that there are at least some philosopher ants that looked up now and again and wondered at their circumstance or marveled at the properties of their world that might provide clues about an underlying reality and their place in it.

I realize that juxtaposing our human response to the universe with the behavior and psychology of ants may not provide the most flattering analogy, but keeping our arrogance in check has value. We have a lot in common with these ants (six legs and lack of lungs notwithstanding). Much like the ant farm inhabitants, we are effectively confined to a limited domain, with restricted access to the cosmos. One could, of course, argue that we also have a key difference—that our human intelligence is capable of trying to understand the universe we are part of (or so we would like to think). We have learned how to extend our awareness of reality beyond that of our own senses, using sophisticated equipment (by contemporary human standards) to probe the universe outside of the ability of our bodies to perceive it. That is a big deal. However, despite all our supposed intelligence, we do a remarkable number of stupid things. I conjecture that at least part of our focus on tempting short-term societal "marshmallows" and our ready willingness to devote our lives to building up piles of crumbs is simple unawareness of how phenomenal the universe is and how special our role might be in it—if we can get out of our technological adolescence unscathed.

As a thought experiment, I sometimes imagine a scenario in which the universe was created as a homework project with the sole purpose of finding out whether sentient life could evolve to understand reality. Sentient life arose in this experiment, but this life decided that scrambling to hoard crumbs is how they want to spend their ephemeral existence. The resulting universe in this experiment becomes akin to a grid in Conway's Game of Life (from Chapter 9) with nothing but blinkers. I don't know about you, but when I'm playing around with the Game of Life and end up with nothing but stagnant forms like blinkers and blocks, I pull the plug and start over.

An Antidote

Of course, ignoring the universe isn't our only possible response to facing the existential. Why did you pick up this book and decide to spend your valuable time reading it? People who read astrophysics books are surely drawn to the topic for diverse reasons—perhaps seeking spectacular images of the universe or looking for answers about how things work (in which cases, this book was surely a disappointment on both fronts). But my suspicion is that underlying these reasons is a thirst to experience awe and be filled with wonder. My belief is that the universe simultaneously causes us to feel part of something bigger than ourselves, invokes humility, and induces awe. This belief is why I decided to write this book.

Exploring the profound mysteries of the universe is a quest that belongs to all of us. Moreover, if we don't find a way to share the wonder of science with all of humanity, we are setting ourselves in opposition to the universe, and we will not fare well. I worry that as our hypotheses become more advanced, the math more complex, and the outcomes more challenging to visualize, it will become increasingly challenging for nonexperts to penetrate the jargon. As the jargon begins to sound incoherent to an outsider, we also risk the jargon being indistinguishable from nonsense. If cutting-edge science and nonsense are indistinguishable to most people, we have a serious problem, ripe for science denial and ready for pseudoscience to make even deeper inroads.

With three kids of my own, in addition to teaching in the trenches for decades, I've watched how science education unfolds on a daily basis. On one hand, schools (at least the public ones) are under a tremendous strain and often don't have the resources to do much beyond following standard established curricula, which comes with set worksheets, homework, and "right" answers. On the other hand, as a college professor I see student after student come through the system as if it is a race to get a job their parents will approve of and that will get them a big pile of crumbs. The fastest (and easiest) route to this end is to follow the *known* path. It strikes me that there is very little room left for deep thinking and creative exploration on this route. Sometimes I worry that our education system has excelled at training creativity out of us, and we learn (both overtly and through inference) that there are established ways of thinking about things from which we should not veer. Scratch the word "sometimes," I worry about this every day.

My own (very strongly held) position is that our world desperately needs more people who learn and think about lots of (apparently) unrelated things. Yes, of course, we need people to go deep and narrow and drill down on very specific questions, but if too many people take this route, we lose the connective tissue, cross-pollination, and innovation that come from a truly broad education. If you are reading this, and there happens to be any children in your life, please take some time to talk with them about whatever topics in this book you found the most intriguing. I hope they will have lots of questions. It is totally OK to answer with "I don't know, but maybe you could be the person to figure that out." Celebrate and nurture their curiosity. I give it even odds that there is a child right now who could think about black holes, or dimensions, or dark energy in some creative and novel way that would open the field and move our understanding forward.

To be sure, we understand a tiny fraction of reality (if that), but the mere fact that we understand that we don't understand is nontrivial.[4] I like to envision reality as a vast landscape, in which we have charted many trails through various terrains, some of which have led to mountain peaks from which we can see how different local ecosystems are connected. Other trails have led to sheer cliff walls we don't yet have the know-how to scale or swamps in which we are easily mired. Still other locations in the landscape can be accessed from multiple trails that begin at distinctly different trailheads. If we want to see what is on the other side of the swamp or at the top of the cliff, we have to keep exploring and—critically—we must be open to new routes that may take us in unexpected directions. It is equally essential that people who have walked different trails convene to share what they learned on their route. If we all walk the same trail, over and over again, we will have deep expertise of that trail, but we stand little chance of gaining significant new insight into the full underlying structure of the landscape.

While we may not be able to see more than a few turtles down, we can also turn our gaze upward to appreciate the view of reality that we already have. Even this view within the horizons of our knowledge is spectacular, and those very horizons are ever expanding. No single human can become an expert in more than a sliver of our collective and cumulative learning. You could spend your entire life exploring the diverse landscape of thought. Given the knowledge that we do have, how do we make the best use of it going into the future?

Homework

Surely you didn't think you could read an entire book written by a professor and not have homework. For better or worse, you don't have to turn anything in. But if you do the homework, it will make me happy, and it might make you happy, too.

1. Teach someone something you learned in this book. I've come to believe that learning is one of the very most important things we humans can do. Which makes teaching like the next most important thing. I have this naively optimistic notion that if more people had a visceral understanding of the vastness of the universe, the world would be a better place. So, humor me, OK?

2. Visit a truly dark night sky and bask in the universe. Depending on where you live, this might be a challenge,[5] but do what you can. Pack a blanket to lie on (you'll regret it if you don't), appropriate clothes, bug repellent, and so forth so that you will be comfortable. You do not want to be distracted by a mosquito or shivering from the cold while trying to contemplate the vastness of the universe. Or maybe you do. You do you. Also, bring a friend or two.

3. Make a long-term habit of thinking about our place in the universe. After you've closed this book and put it on the shelf, it will be easy to go about life ignoring the universe. I don't know if the universe cares, but I do. Find a daily or weekly occasion that you let yourself sink into the existential and how profound it is for us to be alive at all.

Finally, a sincere thank you for taking the time to read this book. I hope that your curiosity is piqued, and you see the universe in a different way than you did about three hundred and fifty pages ago.

Acknowledgments

This is a book that I've started writing three times over the last fifteen years. Amid my waffling and fits and starts, many people have been entrained in the process to whom I owe an enormous debt of gratitude.

First and foremost, my family. I had three small children when I first started playing around with this book. It turns out that having three small children is not overly compatible with finding time to write a book, so I put the project aside and convinced myself that I had no business writing a book anyway. Over the last year, after I picked up this project for the third and final (?) time, my children and husband, Rémy, have provided support and encouragement in myriad ways both large and small and impossible to enumerate. Simply put, this book would not have been written without them.

In the intervening years, hundreds of students have put up with me experimenting on them in class with new material and new ways of presenting it. To be sure, not all experiments succeeded. I have long suspected that when I am teaching, I end up learning even more than the students. So, a big shout-out to my amazing students who have helped me become a better communicator, often through trial and error. Over the last year, several students have given direct feedback on drafts of the manuscript for this book, which has been invaluable. In particular, Jessica Chung has provided constructive criticism on more chapters of this book than probably anyone other than my editor.

The second time I picked up this project, my one-time postdoctoral fellow (and now collaborator) Allie Costa agreed to help me turn my crazy idea into a textbook. Allie is a gifted teacher with an innate drive to help improve science education, and the perfect partner for such an endeavor. We spent a lot of time developing an outline and plan for that project. Then COVID hit, and as with many things in the world, that project got derailed. Who knows, maybe we will pick it back up again.

My dear friend Bree Luck provided essential input to help make this book more accessible to people without a deep science background. As an artist and theater director, Bree read this manuscript with an entirely different perspective than anyone else. Her feedback was brilliant and thoughtful, and I owe her a very nice dinner.

I also owe deep thanks to my colleagues in Religious Studies who have graciously put up with me and my naive questions and lines of inquiry. We've had fantastic discussions over the years (at least I think so, and I hope they do, too), which have helped to broaden my perspective and understanding in ways that were important for this book. In particular, I would like to thank Martien Halvorson-Taylor, Kurtis Schaeffer, Willis Jenkins, Matt Hedstrom, Janet Spitler, and Chuck Mathewes.

The third time I picked up this book, it finally stuck in large part thanks to my terrific agent, Jim Levine. Jim is a natural matchmaker for authors and editors, and he connected me with the best possible editor for this book, T. J. Kelleher. I've found a kindred spirit in T.J., and the process of bringing this book to completion has been a joy with his support, guidance, and advice.

Further Reading

CHAPTER 2

Principia Mathematica (2nd ed., 1925)
I don't know how many people may actually want to read *Principia Mathematica* by Whitehead and Russell, but if you are curious, it is worth looking at just for the aesthetic value. It is brimming with cryptic symbols and fun mathematical logic.

Knowledge: A Very Short Introduction (2014)
Want to learn a little more about epistemology, but don't have a lot of time? This little book by Jennifer Nagel is a good place to start.

Perception: A Very Short Introduction (2017)
In the same series as the book listed above on epistemology, this little book by Brian Rogers is short and sweet, providing a general overview of some of the major topics in perception.

The Structure of Scientific Revolutions (1962)
This is an absolute classic on the philosophy of science, scientific progress, innovation, and how we understand the world scientifically. This is not light reading, but the paradigm establishing nature in this text by Thomas Kuhn makes it worth the read.

Thinking, Fast and Slow (2011)

This book by Daniel Kahneman made a big splash when it debuted, and it changed the way I think about thinking. This book is intended for a general audience (as opposed to academics), which makes it eminently readable.

The Amazing Dr. Ransom's Bestiary of Adorable Fallacies (2017)

I bet you didn't expect a book on fallacies to have "adorable" in the title. This book by Douglas Wilson and N. D. Wilson is a gem—utterly whimsical and delightful. The text covers all of the major fallacies, while feeling a bit like a fantasy bestiary.

The Demon-Haunted World: Science as a Candle in the Dark (1996)

This book is classic Carl Sagan, cutting straight to the core of why pseudoscience is dangerous and the perils we face as a scientifically illiterate society. I feel so strongly about this book that I frequently assign it in one of my courses.

CHAPTER 3

The Cosmic Perspective (9th ed., 2019)

This is my favorite introductory astronomy textbook; the lead author, Jeff Bennett, is great at helping the reader build an understanding of core concepts in astrophysics. This book can provide background on many of the topics that come up in this book that you might want to learn more about.

The Inflationary Universe: The Quest for a New Theory of Cosmic Origins (1997)

If you want to take a deep dive into how inflation theory works and learn more about things like "false vacuums," this book by one of the theory's founders, Alan Guth, is a great place to start.

The Life of the Cosmos (1997)

From the title alone you may not infer that, at its core, this book is about the implications of our own universe being inside a black hole. In this book Lee Smolin lays out the vast ramifications of black hole cosmology, and what this could mean for the laws of physics.

Warped Passages: Unraveling the Mysteries of the Universe's Hidden Dimensions (2005)

If you want to go down the rabbit hole, this book is a good entry point. The author, Lisa Randall, is one of the architects of brane theory, bringing together gravity, string theory, and extra dimensions, large and small.

The First Three Minutes: A Modern View of the Origin of the Universe (2nd ed., 1993)

This is a classic popular book that gives an in-depth overview of the very early universe. For better or worse, despite being a few decades old, our understanding has not substantively changed since Steven Weinberg published this book.

CHAPTER 4

Life in the Universe (5th ed., 2022)

This is the preeminent textbook on astrobiology, led by author Jeff Bennett, who has a deep understanding of pedagogy and learning. This book isn't cheap, but if you really want to learn the material, this is the book you want.

Astrobiology, Discovery, and Societal Impact (2020)

There are a host of books out there on astrobiology, and I'm sure many of them are great. I specifically include this suggestion because the author, Steven Dick, goes beyond the science to discuss the societal factors at play. As a former chief historian for NASA, he has the chops to do this.

What Is Life? (1944, reprint 2012)

This is the classic book by the late Nobel laureate and founder of quantum mechanics Erwin Schrödinger. He had a brilliant mind, and this book (written for a general audience) gives us a glimpse into his thinking.

Astrobiology: A Very Short Introduction (2014)

If you don't have a lot of time but want a quick primer, this is the book for you. David Catling gives a great overview and history of the topic.

CHAPTER 5

Dark Matter and Dark Energy: The Hidden 95% of the Universe (2019)

This is a great readable dive into everything from Chapter 5. The author, Brian Clegg, has a broad portfolio, which gives him a perspective tuned for a general audience.

The 4 Percent Universe: Dark Matter, Dark Energy, and the Race to Discover the Rest of Reality (2011)

Richard Panek takes us into the science behind the scenes as dark energy was discovered. This book will give you a glimpse into the world of cutting-edge science and how the discovery of dark energy unfolded.

The Elephant in the Universe: Our Hundred-Year Search for Dark Matter (2022)

In this text Govert Schilling gets deep into dark matter (without dark energy taking center stage) and our quest to understand what is going on in the universe.

The End of Everything (Astrophysically Speaking) (2021)

While this book isn't specifically on dark matter or dark energy per se, I am impelled to include it just because Katie Mack is brilliant and hilarious. Here Mack takes the reader through the different scenarios in which the universe might end.

CHAPTER 6

Relativity Visualized (1985)

I know this is from 1985, but this book by Lewis Carroll Epstein is my all-time favorite for getting a more intuitive understanding of relativity. I am thrilled and grateful that it is still in print, so please go buy a copy while you can.

Black Hole Survival Guide (2020)

This is a gorgeous book in both text and illustration on black holes. Janna Levin is gifted at making science accessible in her writing, and this book is no exception.

The Information Paradox: A Pedagogical Introduction (2011)

This is a freely available PDF written by Samir Mathur, which you can find at https://arxiv.org/abs/0909.1038. Be warned, this is not for a novice reader and depends on some previous physics background. That being said, it is a deep and thorough discussion of the "paradox."

The Black Hole War (2009)

Leonard Susskind went head-to-head with Stephen Hawking in their battle to understand if information is lost in black holes. Being written by Susskind, you should not be surprised that the book is heavily in favor of string theory and holograms.

Black Holes: A Very Short Introduction (2015)

This book is like the Cliffs Notes for black holes, coming in at less than one hundred pages. Katherine Blundell does a lovely job of distilling the essential physics of black holes into a short, readable format.

CHAPTER 7

The Order of Time (2019)

This is a beautifully written book by Carlo Rovelli that gives accessible overviews of the physics surrounding the nature of time. I have read many books on time, and this is probably my favorite.

Time: A Very Short Introduction (2022)

Like the other books in the Very Short Introduction series, this book by Jenann Ismael is essentially the Cliffs Notes on the nature of time. Written by a philosopher, you will get a slightly different perspective than the books written by physicists.

Time Reborn (2013)

Lee Smolin, as always, is a bit of a maverick, for which I have deep respect. In this book he throws down the gauntlet to physics on the nature of time. If you have read his other work, much in this book will seem familiar.

About Time: Einstein's Unfinished Revolution (1996)

This is the oldest book on the list for this chapter, but, for better or worse, our understanding of time has not really changed in the last quarter century. I'm just an all-around fan of Paul Davies and his writing, so I had to include this one.

CHAPTER 8

Our Mathematical Universe (2015)

Max Tegmark has no qualms about thinking deeply about ideas that might seem outlandish on the surface. In this book, he gets deep into the motivation behind multiple universe conjectures and their mathematical basis.

The Elegant Universe: Superstrings, Hidden Dimensions, and the Quest for the Ultimate Theory (1999)

This is the book that inspired the *NOVA* miniseries by the same name. This book is classic Brian Greene, providing a great accessible introduction to the beauty of string theory.

Warped Passages: Unraveling the Mysteries of the Universe's Hidden Dimensions (2006)

(This book is also recommended for further reading for Chapter 3.) Lisa Randall is one of the architects of brane theory, and this text distills the salient concepts for a general reader and illuminates the thought process of one of the leaders in the field.

CHAPTER 9

Superintelligence: Paths, Dangers, Strategies (2016)

This book will take you on a journey down the rabbit hole to where artificial intelligence may be headed. Guided by Nick

Bostrom, one of the world's preeminent thinkers on the intersection of philosophy and artificial intelligence.

The Road to Reality: A Complete Guide to the Laws of the Universe (2007)
This book is a full five-course meal, so you will need time to digest. That being said, Nobel laureate Roger Penrose goes deep into the math and physics underlying what we know of the universe, with a dash of philosophy thrown in for good measure.

The Lightness of Being (2008)
Nobel laureate Frank Wilczek is one of the world's top experts in particle physics, and with this book he explores that field and the unification of forces (not to mention how ridiculous gravity is).

Gödel, Escher, Bach: An Eternal Golden Braid (1979)
Wow, I can't believe this book is over forty years old. I devoured it in my early twenties, and Douglas Hofstadter's weaving together of aesthetics, physics, and consciousness was exactly what I needed. The main concepts in this book are evergreen.

CHAPTER 10

Just Six Numbers (2001)
Martin Rees is one of the luminaries in the field and has been a prolific author of popular science books. This short book gets deep into some of the most fundamental parameters in the universe.

The Goldilocks Enigma (2008)
I am a big fan of Paul Davies books and his hybrid approach through the lenses of science and philosophy. Read this book side by side with *Just Six Numbers* to get an interesting sense of how Rees and Davies perceive things differently.

Fine-Tuning in the Physical Universe (2020)
This is an edited compendium on several aspects of fine-tuning. Be warned, though: it is meant as a textbook, and it is correspondingly dense and full of high-level concepts.

Large Number Coincidences and the Anthropic Principle in Cosmology (1973)
(In *Proceedings of Symposium: Confrontation of Cosmological Theories with Observational Data*.) I wouldn't normally recommend conference proceedings, but this is the classic paper by Brandon Carter that lays out the anthropic principle in cosmology. You can access this freely online.

Notes

CHAPTER 1: A LITTLE PERSPECTIVE

1. Since 2015, we have the exciting new realm of observations using gravitational waves! But that's for a later chapter.

2. In other words, the light has traveled 13 billion light-years. To be clear, I know that "light-year" sounds like a unit of time (it even has the word "year" in it!), but don't be fooled. A light-year is simply the distance light travels in a year, which is a long way. Given the enormously sized scales involved in talking about the universe, it is helpful to have a unit that can keep up. If we measured everything in miles, or kilometers, that would be silly. For example, just to get to the Andromeda Galaxy—the nearest spiral galaxy similar to the Milky Way, we would need 15,000,000,000,000,000,000 miles. Do you even know how to say that number? (OK, fine, it is 15 quintillion miles.)

3. A. Chirico et al., "Designing Awe in Virtual Reality: An Experimental Study," *Frontiers in Psychology*, 2018; C. Anderson et al., "Are Awe-Prone People More Curious? The Relationship Between Dispositional Awe, Curiosity, and Academic Outcomes," *Journal of Personality*, 2020; V. Griskevicius and M. N. S. Shiota, "Influence of Different Positive Emotions on Persuasion Processing: A Functional Evolutionary Approach," *Emotion*, 2010.

4. D. Keltner and J. Haidt, "Approaching Awe, a Moral, Spiritual, and Aesthetic Emotion," *Cognition and Emotion*, 2003.

5. Keltner and Haidt, "Approaching Awe, a Moral, Spiritual, and Aesthetic Emotion"; J. Piaget and B. Inhelder, *The Psychology of the Child*, Basic Books, 1969.

6. Ironically, the "truth" is that Nietzsche did not actually say this. Instead, he said, "No, facts are precisely what is lacking; all that exists consists of interpretations" (F. Nietzsche, *The Will to Power: An Attempted Transvaluation of All Values*, translated by Anthony Ludovici, T. N. Foulis, 2016 [1913]. Made available through Marc D'Hooghe Free Literature/Project Gutenberg, https://www .gutenberg.org/ebooks/52915). Still, the made-up quote is a good one and clearly has staying power on the internet.

7.   A. C. Clarke, "Hazards of Prophesy: The Failure of the Imagination," in *Profiles of the Future: An Enquiry into the Limits of the Possible*, Harper & Row, 1962.

CHAPTER 2: WHAT IS KNOWLEDGE?

1.   A. N. Whitehead and B. Russell, *Principia Mathematica*, 2nd ed., Cambridge University Press, 1950.

2.   Beware, the simple statement "*I think, therefore I am*" gets us caught in a paradox of self-reference. In this case, for example, by virtue of being an action, "thinking" necessarily changes the state of the subject (you), which renders the definition of who "you" are obsolete by the time you have done the thinking.

3.   You might note the "typically" in the previous sentence—many people have some level of color blindness, and you may well be one of them.

4.   Be warned: there are lots of online tests to determine if you might be a tetrachromat that are pretty much BS. Remember all the stuff of skepticism? Keep that in play.

5.   R. Feynman, *Cargo Cult Science*, Caltech Commencement Address, 1974.

6.   Aristotle facing zombie cats could be a solid plot device for a B-grade movie.

7.   Full disclosure: Feynman gets a lot of credit for this quote, and it is a great sound bite (which is why I deployed it here), but it's not clear that he ever said precisely this. He did certainly have similar statements (e.g., "On the other hand, I can safely say that nobody understands quantum mechanics" (R. Feynman, *The Character of Physical Law*, MIT Press, 1965).

8.   Scientists can have a whole range of things named after them: laws, theorems, physical constants, units, and other normal things. But they can also have odd things like instabilities (e.g., the Parker instability) and holes (e.g., the Lockman Hole) take on their name. That seems like a lot more fun to me.

9.   J. L. Lauritsen and N. White, *Seasonal Patterns in Criminal Victimization Trends*, Bureau of Justice Statistics, 2014; Federal Reserve Economic Data (FRED), *Industrial Production: Manufacturing: Non-Durable Goods: Ice Cream and Frozen Dessert*, https://fred.stlouisfed.org/series/IPN31152N, accessed May 2023.

10.   David Dunning and Justin Kruger have done some interesting work in this area, after whom the "Dunning-Kruger effect" is named, based on their 1999 study with the fabulous title "Unskilled and Unaware of It: How Difficulties in Recognizing One's Own Incompetence Lead to Inflated Self-Assessments" (*Journal of Personality and Social Psychology*, 1999).

CHAPTER 3: WHAT CAUSED THE BIG BANG?

1.   If you want to know more about any of these topics, like comets, molecular clouds, nucleosynthesis, and stellar evolution, I provide some book recommendations in Further Reading.

2.   T. Aquinas, *Summa Theologiae: Questions on God*, Cambridge University Press, 2006.

3.   Augustine, *St. Augustine's Confessions* (translated by E. B. Pusey), Project Gutenberg, 2001. As with many infamous quotes, this supposed quote from Augustine is also not accurate, and, in fact, Augustine says he would not answer this way. In Chapter XI of Augustine's *Confessions*, he writes, "See, I answer him that asketh, 'What did God before He made heaven and earth?' I answer not as one is said to have done merrily (eluding the pressure of the question), 'He was preparing hell (saith he) for pryers into mysteries.'"

4.   E. Kolb and M. Turner, *The Early Universe*, Chapman and Hall, 2018.

5.   It is worth noting that from the perspective of a particle in this *very* early universe, 10^{-32} seconds is a significant fraction of the age of the universe. Moreover, if the particles were traveling at close to the speed of light ("relativistically"), they would have experienced time dilation, and 10^{-32} seconds could be a very long time in their view.

6.   To be fair, there is some debate on this, and Hoyle denied having a pejorative intent. Given his immense dislike for the idea and predisposition to fighting about it, I find myself skeptical.

7.   H. Kragh, "What's in a Name: History and Meanings of the Term 'Big Bang,'" 2013, accessed June 15, 2023, https://arxiv.org/abs/1301.0219.

8.   In all fairness, I should note that there is an alternate hypothesis floating around—the ekpyrotic scenario, which can explain many of the same problems as the standard inflation model. I will talk about this scenario later in the chapter.

9.   C. Indicopleustes and J. W. McCrindle, *Christian Topography of Cosmas, an Egyptian Monk*, Cambridge University Press, 2010.

10.   E. R. Harrison, *Darkness at Night: A Riddle of the Universe*, Harvard University Press, 1987.

11.   L. D. Landau and E. M. Lifshitz, *Statistical Physics*, Pergamon Press, 1958.

12.   E. Hubble, "A Relation Between Distance and Radial Velocity Among Extragalactic Nebula," *Proceedings of the National Academy of Sciences*, 1929.

13.   In what passes for humor among scientists, one of the first papers to detail the calculations involved in Big Bang nucleosynthesis is known as the $\alpha\beta\gamma$ (alpha-beta-gamma) paper. Ralph Alpher did the work with his thesis advisor

George Gamow, and then added Hans Bethe to the author list so the paper could be listed as "Alpher-Bethe-Gamow," or cheekily "αβγ."

14. The horizon of the observable universe is limited by how far light emitted from a region could have traveled since the Big Bang.

15. In case any actual cosmologists ever read this text, which strikes me as extremely unlikely, I want to make sure I point out a subtlety here—when the light from the current horizon was emitted and headed in our direction, the source of that light was roughly 13.7 billion light-years away. However, the universe has been expanding during those intervening years while the light took this road trip. As a result, the source of that light is now actually more than 40 billion light-years away.

16. T. Aquinas, *Summa Theologiae: Questions on God*, Cambridge University Press, 2006.

17. Once again noting that temporal words like "precede" are troublesome here, as time may not have existed.

18. M. Juergensmeyer and W. C. Roof, *Encyclopedia of Global Religion*, SAGE Publications, 2001.

19. W. Wallace, "Empedocles," *Encyclopedia Britannica*, 1911. Empedocles strikes me as an extraordinarily interesting character. If you have a little time to kill, do a little research on him—and in particular stories of his death.

20. J. B. Hartle and S. W. Hawking, "Wave Function of the Universe," *Physical Review D*, 1983.

21. J. Khoury, B. Overt, P. Steinhardt, and N. Turok, "The Ekpyrotic Universe: Colliding Branes and the Origin of the Hot Big Bang," *Physical Review D*, 2001.

22. Once again, several different topics in this book collide here—first the Big Bang and black holes, then dark energy, and now whether the universe is fine-tuned, and—as a bonus—whether there are other dimensions, as necessitated by string theory. Sorry about that, but I don't make the rules here.

23. L. Smolin, *The Life of the Cosmos*, Oxford University Press, 1997.

CHAPTER 4: DOES EXTRATERRESTRIAL LIFE EXIST?

1. Just for fun, you might be interested to know how *Deinococcus radiodurans* was discovered. In the 1950s folks were trying to determine how to best sterilize canned food. They bathed some canned meat in high-energy radiation, which was supposed to kill basically anything. But a can of meat went bad anyway, and they thus made the acquaintance of this "dreadful berry that withstands radiation."

2. Boltzmann even has a physical constant named after him (the Boltzmann constant, k_B), which is perhaps even higher acclaim (and certainly rarer) than winning a Nobel Prize.

3. This could have happened on Earth. We have no way of knowing whether life emerged only to be immediately snuffed out during the Late Heavy Bombardment period. Heck, it could have happened multiple times.

4. As always, being mindful that taking humans as an example is narrow-minded. But it is also the only example we have.

5. Yes, you read that correctly—radio *light*. There is a common misconception about radio waves being audible—since we *listen* to our *radios* you can be forgiven for having this misunderstanding. What these devices we call "radios" actually do is take radio *light* waves and convert them into *sound* waves that we can hear.

6. Because the 21-cm line is so important for astronomy, it is one of only a few small slices of the electromagnetic spectrum that are "protected," and use of this frequency on Earth is forbidden. Although that doesn't mean that it isn't used illegally.

7. Excluding the grim and dystopian possibility that technologically advanced life could kill off any other life-forms.

8. A great set of references on this can be found in Kwon et al., "How Will We React to the Discovery of Extraterrestrial Life?," *Frontiers in Psychology*, 2018.

CHAPTER 5: WHAT ARE DARK MATTER AND DARK ENERGY?

1. I am not alone in my opinion that Vera Rubin should have won a Nobel Prize for this work. I wrote an op-ed on this in *Scientific American* in 2019 when the prize was awarded to others for related work—after Rubin's death.

2. V. C. Rubin and W. K. J. Ford, "Rotation of the Andromeda Nebula from a Spectroscopic Survey of Emission Regions," *Astrophysical Journal*, 1970.

3. From a really good dark-sky location, you can actually see the Andromeda Galaxy with your own unaided eyes. It is sort of hanging off one of the corners of the Pegasus constellation. In fact, there is a whole Greek drama playing out in the sky with the set of constellations in that region (including Pegasus, Andromeda, Perseus, and Cassiopeia).

4. Objects with gravity cause the paths of light to bend around them. The more massive the object, the more significant the curvature of the light path is. This is referred to as "gravitational lensing." If you happen to have a standard wineglass on hand, you can do a respectable job of mimicking a gravitational lens by looking at how the light from images is warped from passing through the foot of the wineglass.

5. G. Bertone, D. Hooper, and J. Silk, "Particle Dark Matter: Evidence, Candidates and Constraints," *Physics Reports*, 2005.

6.   Astronomers do love their acronyms. I sometimes wonder if the success of a project is inversely correlated with the cleverness of its acronym.

7.   P. Tisserand et al., "Limits on the Macho Content of the Galactic Halo from the EROS-2 Survey of the Magellanic Clouds," *Astronomy and Astrophysics*, 2007.

8.   A. Dar, "Dark Matter and Big Bang Nucleosynthesis," in *Dark Matter in Cosmology*, 1995.

9.   B. Carr and F. Kühnel, "Primordial Black Holes as Dark Matter: Recent Developments," *Annual Review of Nuclear and Particle Science*, 2020.

10.   R. Bernabei et al. (DAMA Collaboration), "First Results from DAMA/ LIBRA and the Combined Results with DAMA/NaI," *European Physical Journal C*, 2008; Z. Ahmed et al. (CDMS Collaboration), "Results from the Final Exposure of the CDMS II Experiment," *Science*, 2009.

11.   F. Wilczek, "Time's (Almost) Reversible Arrow," *Quanta Magazine*, 2016.

12.   The ADMX Collaboration and S. J. Asztalos et al., "A SQUID-Based Microwave Cavity Search for Dark-Matter Axions," *Physical Review Letters*, 2010.

13.   G. Bertone, D. Hooper, and J. Silk, "Particle Dark Matter: Evidence, Candidates and Constraints," *Physics Reports*, 2005.

14.   N. Arkani-Hamed et al., "Manyfold Universe," *Journal of High Energy Physics*, 2001; R. Pease, "Brane New World," *Nature*, 2001; J. Garriga and T. Takahiro, "Gravity in the Randall-Sundrum Brane World," *Physical Review Letters*, 2000.

15.   I. Antoniadis et al., "New Dimensions at a Millimeter to a Fermi and Superstrings at a TeV," *Physics Letters B*, 1998. In academic circles, citations are a fundamental currency. The more your paper is cited by other papers, the better. Citations are a sign that the community takes your idea or work seriously enough to test it, rebuke it, or invoke it. Sometimes I wish we had a system for "negative" citations—meaning "Don't read this paper," without having to cite the paper.

16.   G. Landsberg, "Searches for Extra Spatial Dimensions with the CMS Detector at the LHC," *Modern Physics Letters A*, 2015.

17.   I. Banerjee, S. Chakraborty, and S. Sengupta, "Looking for Extra Dimensions in the Observed Quasi-Periodic Oscillations of Black Holes," *Journal of Cosmology and Astroparticle Physics*, 2021; S. Vagnozzi and L. Visinelli, "Hunting for Extra Dimensions in the Shadow of M87*," *Physical Review D*, 2019.

18.   K. Agashe, M. Ekhterachian, and D. Kim, "LHC Signals for KK Graviton from an Extended Warped Extra Dimension," *Journal of High Energy Physics*, 2020; E. Boos, V. E. Bunichev, and I. P. Volobuev, "Heavy Scalar Boson in View of the Unconfirmed 750 GeV LHC Diphoton Excess," *Journal of Experimental*

and Theoretical Physics, 2017; K. Ghosh et al., "Universal Extra Dimension Models with Gravity Mediated Decays After LHC Run II Data," *Physics Letters B*, 2019.

19. B. Betz et al., "Mini Black Holes at the LHC," in *22nd Winter Workshop on Nuclear Dynamics*, 2006.

20. C. Xue et al., "Precision Measurement of the Newtonian Gravitational Constant," *National Science Review*, 2020.

21. D. Clowe et al., "A Direct Empirical Proof of the Existence of Dark Matter," *Astrophysical Journal Letters*, 2006.

22. K. Croswell, "The Constant Hubble War: A 20 Year Fight over the Age of the Universe," *New Scientist*, 1993.

23. Megaparsec is a unit of length that corresponds to a bit more than 3 million light-years. That seems like a long way on our human scales, but in the scheme of the universe, one megaparsec is in our backyard.

24. The words "about" in the previous paragraph to preface the current value of H_0 and "primary" in this sentence are very intentional; we can also use measurements of the cosmic microwave background to determine H_0 independently. At the moment, there is a bit of tension between the two methods, which could be telling us something new and important about the universe. There is a lot of good physics to be had in that discussion, but I don't want to derail us from getting to the discovery of dark energy.

25. There are three different types of redshifts in astronomy: gravitational, cosmological, and relativistic. Galaxies speeding away from us gives rise to relativistic redshift, in which wavelengths of light get stretched out (i.e., reddened) due to the motion of the source that emits them.

26. A. Riess et al., "Observational Evidence from Supernovae for an Accelerating Universe and a Cosmological Constant," *Astronomical Journal*, 1998; S. Perlmutter et al., "Measurements of Omega and Lambda from 42 High Redshift Supernovae," *Astrophysical Journal*, 1999.

27. M. P. Hobson, G. P. Efstathiou, and A. N. Lasenby, *General Relativity: An Introduction for Physicists*, Cambridge University Press, 2014.

28. J. Martin, "Everything You Always Wanted to Know About the Cosmological Constant Problem (but Were Afraid to Ask)," *Comptes Rendus Physique*, 2012.

29. P. Agrawal, G. Obied, P. Steinhardt, and C. Vafa, "On the Cosmological Implications of the String Swampland," *Physics Letters B*, 2018.

30. P. Steinhardt and N. Turok, "A Cyclic Model of the Universe," *Science*, 2002; A. Khodam-Mohammadi, "Unifying Turnaround-Bounce Cosmology in a Cyclic Universe Considering a Running Vacuum Model," *Modern Physics Letters A*, 2022; J. Farnes, "A Unifying Theory of Dark Energy and Dark Matter:

Negative Masses and Matter Creation Within a Modified ΛCDM Framework," *Astronomy and Astrophysics*, 2018.

CHAPTER 6: WHAT HAPPENS INSIDE BLACK HOLES?

1. I have serious cognitive dissonance about warning the reader about upcoming math. I worry that the warning itself plays a role in training people to fear math as something they should be warned about. On the other hand, if you are already afraid of math, I want you know that it will be OK. Please don't be afraid of equations—they are just sentences with very precise symbols and grammar. And sometimes they are essential for helping us build intuition about things we can't (and probably don't want to) experience.

2. To be fair, the notion of "empty" space when we consider quantum mechanical probabilities gets a bit messy. I'm overtly sweeping that under the rug for now.

3. It turns out they may not actually be "black" in the truest sense, but we will come back to that later. For all practical purposes, you can go on thinking of them as "black" for the time being.

4. There is a subtle nuance here to be aware of—light can *try to* escape from the boundary of a black hole, but in doing so it loses all its energy, and in some sense ceases to exist. The semantics here are tricky.

5. In fairness, for people who really care about proper physics, the shape of the object does matter at some level, but let's just assume we are in a world of spherical cows, OK? The precise shapes don't really matter for our purposes here. And I really don't think you want to start doing the necessary calculus. At least I don't.

6. The type of black hole we are dealing with here—a Schwarzschild black hole—is the simplest and most straightforward to understand. There are three additional black hole types that involve rotation and charge, but my editor reminds me that this isn't supposed to be a *whole book* on black holes, so I'm not going to go down those paths. For now.

7. I just can't not take a moment to revel in the name "Schwarzschild." This term gets its name from Karl Schwarzschild, who came up with this solution to Einstein's field equations during World War I—while on the Russian front. (H. Hornung, *The Mystery of the Dark Bodies*, Max Planck Gesellschaft, 2019, https://www.mpg.de/11225504/the-mystery-of-the-dark-bodies). If you happen to know a little German, you may also have noted that "Schwartzschild" translates to "Black Shield," which seems spectacularly coincidental.

8. In fairness, there is a slight subtlety here—this amount of "empty" space is what we would infer if things like electrons and protons were tiny little solid objects. However, in the realm of quantum mechanics, they are actually "clouds" of probability, which can fill a lot of space.

Notes

9. Don't worry, even if we did manage to make a black hole in the Large Hadron Collider, we are not in any danger. I will loop back to this later in the chapter.

10. This is a term we actually use. I hope this doesn't ruin the time-honored pasta for you. If you want to be more technical, this general behavior is referred to as "tidal stretching."

11. A. Almheiri, D. Marolf, J. Polchinski, and J. Sully, "Black Holes: Complementarity or Firewalls?," *Journal of High Energy Physics*, 2013.

12. There is a specific, conventional way in which physicists assign each axis of a space-time diagram, which I am going to divert from here. The versions I'm using here are equally valid but take care to note the axes are different than you may have seen elsewhere.

13. We get ourselves in a bit of a tricky situation here with language, which in colloquial usage isn't really set up to talk about "speed" with respect to time. What does that even mean? Mathematically, we think of it simply as a derivative (which will mean something to you if you've had calculus), but that doesn't make it easier to talk about at a dinner party.

14. We owe this fun knowledge to Dr. Persi Diaconis, both a mathematician and magician who led a study on this published in 1990.

15. As a reminder, there are different kinds of "nothing," which we encountered in the chapter on the Big Bang.

16. I know you want to know what constitutes an "observation." So do I. This is actively debated in physics and philosophy, but more or less comes down to a quantum system interfacing with something macroscopic that needs the quantum system to make up its mind.

17. J. Maldacena and L. Susskind, "Cool Horizons for Entangled Black Holes," *Fortschritte der Physik*, 2013. Abstract at Astrophysics Data System, https://ui.adsabs.harvard.edu/abs/2013ForPh..61..781M/abstract.

18. L. Susskind, L. Thorlacius, and J. Uglum, "The Stretched Horizon and Black Hole Complementarity," *Physical Review D*, 1993.

19. J. Baez, "This Week's Finds in Mathematical Physics," John Baez's Stuff (website), 2004, accessed June 2023, https://math.ucr.edu/home/baez/week207 .html.

20. Baez, "This Week's Finds in Mathematical Physics."

21. S. Mathur and O. Lunin, "AdS/CFT Duality and the Black Hole Information Paradox," *Nuclear Physics B*, 2002.

22. S. Haco, S. Hawking, M. Perry, and A. Strominger, "Black Hole Entropy and Soft Hair," *Journal of High Energy Physics*, 2018.

23. Maldacena and Susskind, "Cool Horizons for Entangled Black Holes."

24. Maldacena and Susskind, "Cool Horizons for Entangled Black Holes."

25. A. Chamseddine, V. Mukhanov, and T. Russ, "Black Hole Remnants," *Journal of High Energy Physics*, 2019.

26. P. Chen and R. J. Adler, "Black Hole Remnants and Dark Matter," *Nuclear Physics Proceedings Supplements*, 2003.

27. This is one of those places where I really struggle with the order of the chapters. We will talk more about how space-time can behave this way in the chapters on both the Big Bang and on dark energy. If you want to skip ahead to those and get a quick primer on inflation, this part might make more sense.

28. L. Smolin, "Did the Universe Evolve?," *Classical and Quantum Gravity*, 1992; J. Brockman, "Smolin vs. Susskind: The Anthropic Principle," Edge, 2004, https://www.edge.org/conversation/lee_smolin-leonard_susskind-smolin -vs-susskind-the-anthropic-principle.

29. R. Casadio, A. Kamenshchik, and I. Kuntz, "Covariant Singularities in Quantum Field Theory and Quantum Gravity," *Nuclear Physics B*, 2021.

CHAPTER 7: WHAT IS THE NATURE OF TIME?

1. I mean "event" very generally here—beyond the sense that a party is an event. Rather anything that ever happens is an "event." You reading this note is an "event."

2. This is a real level of commitment because there are four of them; I still think Euler's equation is a better choice.

3. Emmy Noether's story doesn't get nearly enough attention. She was among the first women in the world to get a doctorate in mathematics, and the obstacles and discrimination she dealt with along the way make her breakthroughs even more astounding.

4. I say this with a nod to Descartes and the chapter on epistemology and whether we can actually "know" anything.

5. Admittedly, I have not stopped random people on the street and asked them about whether they thought time was linear, or whether time has loops. My instinct is that such interactions wouldn't end well for me. This may be fertile ground for an aspiring psychology PhD student.

6. Note that we are only considering a single dimension of space to simplify the visualization.

CHAPTER 8: ARE THERE HIDDEN DIMENSIONS?

1. As an example, the word "algebra" comes from the Arabic *al-jabr*, which loosely means "the joining of broken parts."

2. P. Ball, "Islamic Tiles Reveal Sophisticated Maths," *Nature*, 2007.

3. M. Gardner, "Extraordinary Nonperiodic Tiling That Enriches the Theory of Tiles," *Scientific American*, 1968. This seems like a fine place to point

out that there are different sizes of infinities in math. In this case, "uncountable" infinities are larger than "countable" ones.

4. A. C. Clarke, "Hazards of Prophesy: The Failure of the Imagination," in *Profiles of the Future: An Enquiry into the Limits of the Possible*, Harper & Row, 1962.

5. J. Horgan, "Why String Theory Is Still Not Even Wrong," *Scientific American*, 2017; R. Peierls, "Where Pauli Made His 'Wrong' Remark," *Physics Today*, 1992.

6. A. F. Ali, M. Faizal, and K. Mohammed, "Absence of Black Holes at LHC Due to Gravity's Rainbow," *Physics Letters B*, 2015.

7. P. Hořava and E. Witten, "Eleven-Dimensional Supergravity on a Manifold with Boundary," *Journal Nuclear Physics B*, 1996.

8. Abbott et al., "Tests of General Relativity with GW170817," *Physical Review Letters*, 2019.

9. This very book is another case in point; I have to view this book as an experiment, and if I were too afraid to write it because it might not turn out well, I never would have put pen to paper, or fingers to keyboard, as it were.

10. I really hate using the word "never." However, based on the laws of physics as we currently understand them, the word "never" seems appropriate. That being said, I would love for some far-future reader to pick up this book and laugh at how naive and closed-minded we twenty-first-century humans were.

11. M. Tegmark, "Parallel Universes," in *Science and Ultimate Reality: From Quantum to Cosmos*, Cambridge University Press, 2003.

12. A. Guth, "Eternal Inflation and Its Implications," *Journal of Physics A*, 2007.

13. P. Steinhardt, "Natural Inflation," in *The Very Early Universe*, Cambridge University Press, 1983.

14. M. Kramer, "Our Universe May Exist in a Multiverse, Cosmic Inflation Discovery Suggests," Space.com, 2014, https://www.space.com/25100-multiverse-cosmic-inflation-gravitational-waves.html.

15. S. Hawking and T. Hertog, "A Smooth Exit from Eternal Inflation?," *Journal of High Energy Physics*, 2018.

16. "Taming the Multiverse: Stephen Hawking's Final Theory About the Big Bang," University of Cambridge, 2018, accessed June 2023, https://www.cam.ac.uk/research/news/taming-the-multiverse-stephen-hawkings-final-theory-about-the-big-bang.

17. J. Brockman, "Smolin vs. Susskind: The Anthropic Principle"; D. Overbye, "About Those Fearsome Black Holes? Never Mind," *New York Times*, 2004.

18. The idea of branes took off in the late 1990s with a set of papers by Lisa Randall and Raman Sundrum (called the "RS models") and Merab Gogberashvili

(working on a shell model) (L. Randall and R. Sundrum, "A Large Mass Hierarchy from a Small Extra Dimension," *Physical Review Letters*, 1999; M. Gogberashvili, "Hierarchy Problem in the Shell-Universe Model," *International Journal of Modern Physics D*, 2002).

19. V. Rubakov and M. Shaposhnikov, "Do We Live Inside a Domain Wall?," *Physics Letters B*, 1983.

20. J. Khoury et al., "The Ekpyrotic Universe: Colliding Branes and the Origin of the Hot Big Bang," *Physical Review D*, 2001.

21. C. Collaboration, "Search for Microscopic Black Hole Signatures at the Large Hadron Collider," *Physics Letters B*, 2011; L. Visinelli, N. Bolis, and S. Vagnozzi, "Brane-World Extra Dimensions in Light of GW170817," *Physical Review D*, 2018.

CHAPTER 9: WHAT DETERMINES THE LAWS OF NATURE?

1. M. Gardner, "The Fantastic Combinations of John Conway's New Solitaire Game 'Life,'" *Scientific American*, 1970.

2. At the time that I am writing this, if you google "Conway's Game of Life," the results page will have a Game of Life evolving across it, which is a level of nerdiness I am down for. After this quick search, you should have several online options to choose from. These generally have options you can play around with, like the cell size to zoom out or in, the frame rate, and so on. Just experiment and see what happens.

3. I am sorry to say that, although I went to science camp every summer, I never had the chance to attend metaphysical philosophy camp. I wonder if they debate whether s'mores exist.

4. G. Oppy, "Ontological Arguments," *Stanford Encyclopedia of Philosophy*, 2021.

5. A. Franklin, "The Rise and Fall of the 'Fifth Force': Discovery, Pursuit, and Justification in Modern Physics," *American Institute of Physics*, 1993. Indeed, a fifth force was proposed in the 1980s (Fischbach et al., "Reanalysis of the Eoumltvös Experiment," *Physical Review Letters*, 1986), but as researchers went around the empirical inquiry loop, the observations ultimately kicked them onto the path of "Where did our assumptions or method go wrong?"

6. J. C. Maxwell, *A Treatise on Electricity and Magnetism*, Clarendon Press, 1873.

7. The physicist responsible for the levitating frog, Andre Geim, won the Ig Nobel Prize (which is the satiric version of the Nobel Prize) for this in 2000. And then he went on to win the actual Nobel Prize in physics in 2010 for work on graphene. I imagine his lab is a *very* interesting place.

8. In modern physics the term of art converted from "force" to "interaction," which acknowledges that what we perceive as forces really arise from interactions with fields. Frankly, I also happen to think that "interaction" just sounds more friendly and welcoming, so I'm all for it. The strong and weak forces are typically referred to as "interactions," but for the sake of not being any more confusing I'm just calling everything a "force."

9. A quick clarification that protons have not been observed to decay when *in isolation* (outside a nucleus). This is important in the chapter on fine-tuning. Protons *can* decay by capturing electrons.

10. I occasionally give a homework assignment that involves students dropping drops of ink from different heights onto paper and observing the splatter patterns that are created—which have spectacular symmetries in them. This particular assignment invites a lot of questioning looks. You are welcome to try this experiment, which may well spur some curious conversations with passersby. The resulting ink splatters are stunning.

11. Roughly 1 in 10,000 people have their hearts on the right side, which is known as dextrocardia. Still asymmetric, just with the opposite handedness.

12. A. Salam and J. Ward, "A Treatise on Electricity and Magnetism," *Nuovo Cimento*, 1959; S. Weinberg, "A Model of Leptons," *Physical Review Letters*, 1967.

13. Yes, I know, sound requires a medium to travel in, and typically we don't expect there to be sound in outer space given the extremely low density in most of the universe today. But let's not forget that the early universe, at the time of inflation, was a fair bit more dense. In fact, at that time the universe was a thick hot plasma, and there most definitely was sound, and these ancient sound waves are now frozen into the structure of the cosmic microwave background.

14. R. Descartes, *Discourse on Method and Meditations on First Philosophy*, Yale University Press, 1996. With modern-day AI, this also turns out to be wrong, and chatbots are giving all manner of replies.

15. B. Voytek, "Are There Really as Many Neurons in the Human Brain as Stars in the Milky Way?," *Nature*, 2013. There is urban lore that the brain has more nerve cells than stars in our galaxy. I suspect the origin of this myth is ultimately connected to humans wanting to feel important. If you are so inclined, a quick fact check will tell you that the Milky Way has at least 100 billion stars, and possibly as many as 400 billion. Curiously, this is roughly the same number as there are galaxies in the observable universe. Conspiracy theorists must love this.

16. A. Turing, "Computing Machinery and Intelligence," *Mind*, 1950.

17. See, for example, the Illustris simulation and its descendants (Illustris Project [website], accessed June 2022, https://www.illustris-project.org/.)

18. D. A. Grier, *When Computers Were Human*, Princeton University Press, 2005.

19. N. Bostrom, "Are You Living in a Computer Simulation?," *Philosophical Quarterly*, 2003. The original paper by Nick Bostrom is freely available online at https://simulation-argument.com/simulation.pdf.

20. S. Hossenfelder, "The Simulation Hypothesis Is Pseudoscience," *Back-ReAction*, 2021.

21. Far be it from me to throw shade at Descartes, but this issue with self-reference is also embedded in "I think, therefore I am." Thinking is an action that changes the state of being, so after you have done the thinking, you are no longer the same person you were when you started, which brings into question the nature of "am" and what "am" includes.

22. A. Tarski, "The Semantic Conception of Truth: And the Foundations of Semantics," *Philosophy and Phenomenological Research*, 1944; A. Tarski, *Logic, Semantics, Metamathematics: Papers from 1923 to 1938*, Clarendon Press, 1956.

23. P. Raatikainen, "Gödel's Incompleteness Theorems," *Stanford Encyclopedia of Philosophy*, 2020; K. Gödel, *On Formally Undecidable Propositions of Principia Mathematica and Related Systems*, Dover, 1992.

24. S. Hawking, "Godel and the End of Physics," Stephen Hawking (website), 2002, accessed June 2023, https://www.hawking.org.uk/in-words/lectures/godel-and-the-end-of-physics.

25. R. Penrose, "Gödel, the Mind, and the Laws of Physics," in *Kurt Gödel and the Foundations of Mathematics*, Cambridge University Press, 2011.

26. R. Penrose, *The Emperor's New Mind: Concerning Computers, Minds, and the Laws of Physics*, Oxford University Press, 1989.

27. P. Raattkainen, "On the Philosophical Relevance of Godel's Incompleteness Theorems," *Revue internationale de philosophie*, 2005.

CHAPTER 10: IS THE UNIVERSE FINE-TUNED?

1. For a deeper overview of the parameters in the universe, see M. Tegmark et al., "Dimensionless Constants, Cosmology, and Other Dark Matters," *Physical Review D*, 2006.

2. If you don't actually have a good physics book, you should have one on your shelf. You never know when you might need it. If nothing else, they are really good for pressing flowers.

3. G. W. Anderson and D. J. Castaño, "Measures of Fine Tuning," *Physics Letters B*, 1995.

4. G. Hooft, "Naturalness, Chiral Symmetry, and Spontaneous Chiral Symmetry Breaking," in *Recent Developments in Gauge Theories*, NATO Science Series B, 1980.

Notes

5.　In my opinion, this is one of those ridiculous astronomy terms that just don't make sense—for example, there is nothing "hydro" (which suggests water) about this, and "static" (meaning it isn't changing) and "equilibrium" strike me as redundant.

6.　F. Adams, "Stars in Other Universes: Stellar Structure with Different Fundamental Constants," *Journal of Cosmology and Astroparticle Physics*, 2008.

7.　H. Kragh, "When Is a Prediction Anthropic? Fred Hoyle and the 7.65 MeV Carbon Resonance," *Philosophy of Science Archive*, 2010.

8.　Sorry for the jargon. Hopefully, the "triple" clued you in to there being three of something. Those somethings are helium nuclei—so two protons and two neutrons each, which are called "alpha particles." In other words, the triple-alpha process uses three helium nuclei.

9.　F. Hoyle, "On Nuclear Reactions Occurring in Very Hot Stars. I. The Synthesis of Elements from Carbon to Nickel," *Astrophysical Journal Supplement Series*, 1954.

10.　H. Oberhummer, A. Csótó, and H. Schlattl, "Stellar Production Rates of Carbon and Its Abundance in the Universe," *Science*, 2000.

11.　*How* do we do that? You will surely be relieved to know that there are dedicated scientists in labs whose job it is to determine these constants as precisely as possible (for example at the National Institute of Standards and Technology, NIST). You know you want to go read up on these precision lab experiments.

12.　As a throwback to the chapter on laws, you might note that the neutron decay uses the weak interaction.

13.　Just a quick reminder of note 9 in Chapter 9 and the circumstances under which protons can decay.

14.　I don't think I believe in reincarnation, but if I did, I would not want to come back as a face mite.

15.　M. Tegmark, "On the Dimensionality of Spacetime," *Classical and Quantum Gravity*, 1997.

16.　Just to decode the jargon in case you are interested: the Greek letter σ (sigma) is generally used as a statistical measure of difference in a set of measurements (i.e., standard deviation). The 8 comes from this particular measurement being normalized to units of 8 Mpc/h, and h itself is code for $H_0/$ (100 km/s/Mpc).

Sometimes you will also see the letter Q used for the amplitude of primordial fluctuations, which is a straight value of amplitude as observed in the cosmic microwave background and has a value of roughly 10^{-5}.

17.　G. Hinshaw et al., "Nine-Year Wilkinson Microwave Anisotropy Probe (WMAP) Observations: Cosmological Parameter Results," *Astrophysical Journal Supplement Series*, 2013.

18. M. Tegmark, "Why Is the Cosmic Microwave Background Fluctuation Level 10-5?," *Astrophysical Journal*, 1998.

19. R. V. Wagoner, "Big-Bang Nucleosynthesis Revisited," *Astrophysical Journal*, 1973.

20. As a side note, helium is a nonrenewable resource on Earth, and we are constantly losing it to space. Think about that the next time you are holding a helium balloon.

21. S. Rahvar, "Cosmic Initial Conditions for a Habitable Universe," *Monthly Notices of the Royal Astronomical Society*, 2017.

22. M. Tegmark et al., "Dimensionless Constants, Cosmology, and Other Dark Matters," *Physical Review D*, 2006.

23. C. H. Bennett, "Logical Depth and Physical Complexity," in *The Universal Turing Machine—a Half-Century Survey*, Oxford University Press, 1988.

24. Bennett, "Logical Depth and Physical Complexity."

25. Note that the orbits are not actually perfect circles, but back in the day it sure seemed like it.

26. I. Newton, *Opticks: or, A treatise of the reflections, refractions, inflexions and colours of light* (London: Printed for Sam. Smith, and Benj. Walford, 1704), Smithsonian Libraries online, https://library.si.edu/digital-library/book/optickstreatise00newta.

27. C. Darwin, *The Origin of Species*, William Collins, 2011 (1859).

28. B. Carter, "Large Number Coincidences and the Anthropic Principle," in *Confrontation of Cosmological Theories with Observational Data*, D. Reidel, 1973.

29. R. Dicke, "Dirac's Cosmology and Mach's Principle," *Nature*, 1961.

CHAPTER 11: WHAT IS OUR PLACE IN THE UNIVERSE?

1. A. Casselman, "Strange but True: The Largest Organism on Earth Is a Fungus," *Scientific American*, 2007, https://www.scientificamerican.com/article/strange-but-true-largest-organism-is-fungus/.

2. For context, there are other macroscopic living organisms on Earth that have reached ages far surpassing this—including several species of plants believed to be more than ten thousand years old.

3. F. Falchi et al., "The New World Atlas of Artificial Night Sky Brightness," *Science Advances*, 2016, https://www.science.org/doi/10.1126/sciadv.1600377. Don't even get me started on the impending deluge of satellites in Earth orbit. Today as I write this book, global internet providers have submitted applications to launch over half a million satellites into Earth orbit. By comparison, only about five thousand stars are visible to a typical human eye at the darkest locations on Earth. If only a quarter of these planned satellites are ultimately launched, and

only one in ten of those are bright enough for the human eye to see, there will be nearly three times more satellites visible and orbiting in the night sky than visible stars.

4.   I don't know that I can say the same of my dogs; I have the sense that they just accept the nature of reality without wondering why it is the way it is. They certainly don't think about the philosophical implications of quantum mechanics.

5.   There are resources online to help you find the darkest night-sky location near you. One of my favorites is https://www.lightpollutionmap.info, which will let you zoom into any location in the world. Also a pro tip: be mindful of the moon phase—the closer to a new moon, the better.

Index

adaptation, as requirement for life, 84
after this (fallacy), 40
algorithmic compressibility, 323–325
amino acids, and life on Earth, 96–97
amphipod, 82, 82 (fig.)
Andromeda Galaxy, gravitational dynamics, 130, 130 (fig.)
angular momentum, 221
anthropic principle, and fine-tuned universe, 331–334, 337
anti–de Sitter/conformal field theory correspondence (AdS/CFT), as solution for the information paradox, 194–195
antimatter
    asymmetry with matter, 184, 224, 321–323
    and black hole evaporation, 184
    description and characteristics, 184
    mutual annihilation with matter, 224
    pockets of in the universe, 224–225
    tweaking of values, 321–323
antiparticles, 184–186, 219
Aquinas, Thomas, 52, 67
argument from ignorance ("God of the gaps" fallacy), 42–43, 277, 335
Aristotle, 29–30, 204, 302, 333
Aristotle's laws of thought, 29–30

assumptions, 22, 28, 31–33
    See also axioms
astrophysics
    education of author, 2–3
    jargon, 348
    knowledge in, 7, 9
    predictions and observations in, 263–265
    unsolved mysteries, 4
asymmetry
    matter-antimatter, 184, 224, 321–323
    in physics, 224, 279
    and time, 215, 223
    in the universe and humans, 286–287
atmosphere, 94–95, 117–118
atoms
    creation after Big Bang, 63–64
    and strong force, 283, 284
    and weak force, 285–286
awe, as experience, 11–12, 343–344, 347
axioms
    in daily life, 28, 29
    description and truth of, 28, 29 (fig.), 31
    and knowledge, 22, 28
    laws of thought, 29–31
    proof of, 54–55

Index

formation and types, 161–162
and gravity, 156–157, 161–162, 163
hypotheses and testing, 197–198
and light, 159, 162–163
as MACHOs, 133
mass, 156, 161–162, 164
micro–black holes, 251
misconceptions about, 155–156
primordial black holes, 134,
  161–162
radiation emission, 185
and singularities, 161, 198, 199–200
size and density, 156, 159–161,
  164, 195
soft hair of, 196
and space-time warping, 172, 186,
  197–198, 197 (fig.)
and time, 165–168, 176, 202
understanding of, 151, 153–154
BMS symmetries (supertranslation
  symmetries), 196
Bohm, David, 214
Boltzmann brains (spontaneous
  "brains" theory), 99–100, 336
Bonhoeffer, Dietrich, 42–43
brains (human), 16–17, 26–28,
  78–79, 294
brane (multidimensional brane),
  75–76, 253–254
braneworld cosmologies, 261–262
broadcasting civilizations, 102,
  105–107, 108, 114
Broglie, Louis de, 214
bulk, in brane, 75, 262
Bullet Cluster, 138

C-symmetry (charge), 221, 222
Calabi-Yau manifolds, 249–250, 250
  (fig.), 261, 336
carbon, 89, 313–314, 332

carbon dioxide (CO_2), and life on
  Earth, 95
Carter, Brandon, 331
causality, 51, 52, 215
causation, and correlation, 39–40
cause, as idea and types of,
  333–334, 337
cause and effect, decoupling of, 72
cells, 84, 269
chance (or luck), as option for the
  universe being well suited to our
  being here, 334
chaotic inflation hypothesis (eternal
  inflation), 259, 260–261
charge-parity (CP) symmetry, 286
chatbots and AI, 294–295
ChatGPT, 295
chauvinism, as option for the universe
  being well suited to our being
  here, 336
Clarke, Arthur C., 17, 38, 199, 242
Clarke's three laws, 17
CMB (cosmic microwave background)
  light, 63–64, 115
cognitive dissonance, 27–28
colliders, energy in, 289
"color," of gluons and quarks, 284
color receptors, 26
combined symmetry (CPT symmetry),
  222–223
compactified, as term, 247–248
compactified dimensions, 136,
  247–251, 253
complex numbers (imaginary
  numbers), 211–212, 243
computers
  original meaning, 299
  as thinking for simulation,
    294–295, 298–300
  universe as computer, 299–300, 324

Kelsey Johnson is a professor of astronomy at the University of Virginia, former president of the American Astronomical Society, and founder of the award-winning Dark Skies, Bright Kids program. She has won numerous awards for her research, teaching, and promotion of science literacy. She lives in rural Virginia with her family, including two very large dogs.